To my lovely wife of nearly 25 years and our three terrific children
Caitlin (21), Peter (19), and Sam (15)

and to the late James V. Mudge, Jr. (Jamey), 1951–2004, who left this
world far too early

# Preface

This book is about the boys from the high school class of 1969 and their transition to adulthood during the Vietnam-era. I am especially interested in how three prominent pathways from adolescence to adulthood affected the boys' self-concepts. Although there are many avenues to adulthood, and people often combine pathways, I am particularly interested in comparing the context effects that full-time military service (especially in wartime), full-time labor force participation, or full-time college had on a diverse, nationally representative group of American high school boys. The boys in this study grew up and entered adulthood during times unique in American history. From the relative peace and prosperity of post-World War II America, to the dawn of the modern civil rights movement and the escalation of hostilities in Vietnam, to the chaos and bitterness of a nation divided in the late 1960s and early 1970s, and to America's withdrawal from Vietnam and the "stagflation" of the 1970s, these young men experienced more good and bad national and world events in their first two decades of life than most generations in American history.

Make no mistake, their childhood, teenage, and young adult years also afforded plenty of opportunity for fun and hopefulness. They were the first generation brought up on television and drive-in movies. They cruised with friends on Friday and Saturday nights, saw the birth of the Space Age, and were an integral part of the heyday of rock-and-roll. All these pointed to an exciting time brimming with optimism and the smug self-assurance many young people of the day felt toward America's moral, economic, political, and military supremacy. This of course changed in the 1970s as the country seemed to be pulling apart due to such events as Vietnam War,

Watergate and President Nixon's resignation, the advent of forced bussing in pursuit of school desegregation and the ensuing white flight, and the seemingly ever-present domestic terrorists hell-bent on bringing America and its economy to its knees. Groups like the Weather Underground, Black Panthers, and Symbionese Liberation Army dominated many headlines in the early and mid-1970s. In one decade, the country went from a prideful and sometimes ethnocentric nation to one touching on xenocentrism and malaise. The boys in this study were all part of that shift. Just as the young men and women of the post-World War I years are sometimes called The Lost Generation, the boys in my study might aptly be called The Jaded Generation.

The constant threat of World War III and nuclear annihilation hung over everyone's head, creating a constant source of stress and anxiety pervasive in American culture as the boys grew toward adulthood. No schoolchild from the 1950s could easily forget the duck-and-cover drills at school, the fallout shelter signs and rooms filled with water and rations, and the shrill whine of monthly civil defense sirens echoing through the neighborhood. Today they warn of tornadoes, then enemy missiles. When the kids started listening to Top 40 radio, they heard the high pitch signal of the Emergency Broadcast System and the announcement: "This is a test of the Emergency Broadcast System. In the event of an actual emergency . . . ."

Still, for many, the 1950s and 1960s, especially, were idyllic. Children generally felt safe outside the home and starting in kindergarten or first grade, many walked to and from school everyday (and sometimes home for lunch, too). Playing tag or "Statue" or "Pompompullaway" after dark on warm summer nights with all the kids in the neighborhood was a favorite pastime for many. Sonic booms, the invisible though unmistakable sign of the marvels of supersonic flight, thundered across many of the country's cities and farms. When some new milestone in space travel was occurring or a space capsule splashed down somewhere at sea, the nation tuned en masse to live television coverage.

How, you may be asking yourself, do I know all this? Some of my knowledge of American culture and society from the 1950s to the 1970s comes from researching the subject. But in reality, most comes from first-hand experience because I too am part of the birth cohort of 1951. I experienced firsthand many of the events the boys in my study experienced. I joined the Army in the summer after high school, was trained as a para-trooper and served stateside and overseas as a scout in the 82nd Airborne Division. Like millions of other young men, I came home confused and bitter, sheepish about wearing my uniform in public, and reluctant to reveal my veteran status. Many close friends were vehemently opposed to the war (and some the warriors). I was not a fan of the war, but I could not in good conscience stay out of the military while my brothers were fighting and dying. In fact, Vietnam is probably the main reason I went to college on the G.I. Bill and eventually became a sociologist. Still, boys in my Minneapolis city high school (Patrick Henry) entered the United States military in droves, either as volunteers right out of high school or as draftees (usually the year following graduation). Some went to college, but most went to work and waited for their draft notice or joined up to forestall the inevitable and have some choice in their branch and assignment. Some were killed, including a cousin, and others wounded. One high school friend lost an eye and another cousin lost a leg to a landmine. Occasionally the principal would announce over the P.A. system that one of our alumni had been killed and you could hear the wails of stunned and grief-stricken teenagers echo down the halls.

Again, not all was bad. I, like thousands of other young people, hitchhiked across the country with a mere $20 in my pocket. In the spirit of the time, truckers, families, and miscellaneous people on the road treated me to meals, stories, and the pleasure of their company with no thought of reimbursement, save doing the same for someone else if the opportunity arose. I was secure in my large extended family, had tons of cousins to play with, and went to the same high school my parents did. It was quite common for people to live close to where they grew up and for kids to know only one house. For me, however, the reminder of war was everywhere. Most of my friends' fathers, including mine, served in World War II. On Saturday nights during high school my friends and I made a point of getting a case of beer and watching *The Sands of Iwo Jima*, *Guadalcanal Diary, Wake Island*, or *The Flying Leathernecks* whenever they or other war movies appeared on television, which was often. Walking to school and meeting my friends along the way was a welcome routine. And my older sister had lots of cute girlfriends for me to admire from a distance.

North Minneapolis, my neighborhood, was ringed by Victory Memorial Drive on its north and west borders. Every huge elm tree along the expansive parkway of about five miles was planted in memory of a serviceman from Hennepin County, Minnesota, who died during World War I. There were hundreds of trees and at the base of each, a small cross or Star of David was inscribed with his name, rank, and branch of service (many died of diseases, including the influenza pandemic of 1918–1919). Like many of the other boys in my neighborhood, I spent countless hours playing touch football among the trees and faux tombstones or jogging down its well-kept grass. And like most of my friends, I worked during high school. My first job was at a Baskin-Robbins ice cream parlor down the block from my house. Later I worked as a busboy at a locally renowned dinner theater. My best paying job was at a stinky asphalt plant a few miles away from my house, across a massive railroad yard.

The upshot of all this is that I know these data far more intimately than most researchers know theirs. Although I wasn't in the study, I feel like I'm one of the boys in it. I also believe my mostly lower middle class Irish Catholic upbringing was pretty typical of the years the boys in my study were being raised and were entering early adulthood.

## Research Goals

The book has four primary research goals, most examined in chapter-length treatments. The goals are listed in their order of appearance.

1. To correct the sample for selection bias due to attrition over the eight year span of the study, through probit modeling (chapter 2)
2. To develop a context choice schema which predicts entrance into one of the post-high school social contexts, through multinomial logistic regression (chapter 3)
3. To contextualize the early life course experiences of the subjects by constructing a demographic and cultural profile of the high school class of 1969 or birth cohort of 1951 (chapter 3)
4. To examine the role that post-high school social contexts play in adult self-concept development during the transition to adulthood, through the use of structural equation modeling (chapter 4)

## Outline of the Book

Chapter 1 reviews the relevant literature and lays the theoretical and empirical foundations for the book. It sets the stage for the structural equation models appearing in chapter 4. Three broad literatures are reviewed: (1) life course and human development, especially concerning self and attitude change in different phases of life; (2) socialization to role-identities, and (3) occupational choice.

Chapter 2 has three important tasks. First, it describes the Youth in Transition (YIT) data set used in my study of the influence of social context on the self. Briefly, YIT is a five-wave longitudinal study of 2,213 10th-grade boys enrolled in over 80 American public schools in the fall of 1966. The fifth and last wave of data were collected eight years later in the spring of 1974, five years after most had graduated from high school. The study was conducted at the Institute for Social Research at the University of Michigan. The data were obtained through the Inter-university Consortium for Political and Social Research (ICPSR). Second, the chapter presents extensive confirmatory factor analyses of the dimensionality of the self-esteem construct, an important task in establishing the reliability and validity of my key outcome measure. In addition, the self-esteem construct's structural invariance is analyzed and tested. This test is crucial in longitudinal research, although too frequently overlooked. Establishing structural invariance is necessary when change over time is being assessed. Without it, change estimates are fraught with error since one cannot differentiate real change in scores from change due to the nature of the construct itself. Finally, sample selection bias is assessed via probit modeling, per research Objective 1. Accounting for possible differences between those who remained in the study versus those who dropped out along the way is crucial to estimating valid parameter estimates in the structural equation models appearing in chapter 4.

Chapter 3 addresses research Objectives 2 and 3. The second research objective examines issues surrounding post-high school social context choice through the use of multinomial logistic regression. For example, during the five-year period after high school, 17% of the subjects had served on active military duty for at least 18 months or were still in the service, 13% were primarily full-time workers, and 31% had primarily been full-time college students. These are the boys I focus on. An additional 39% were off-and-on full-time workers and college students, or had pursued other activities after high school. The book does not focus on this diverse group of subjects. Research Objective 3 attempts to contextualize the early life course experiences of the subjects introduced in chapter 2. Here I construct a demographic and socio-cultural profile of the birth cohort of 1951 (when the vast majority of the subjects were born). The cohort of young men in my study had many noteworthy life course experiences in their childhood, adolescence, and young adulthood. For example, they went to elementary school when Baby Boom children were flooding the nation's over-extended schools, resulting in typical class sizes of 30–50 pupils. They graduated from high school at the height of the Vietnam War (when over 500,000 U.S. troops were stationed in and around Vietnam); were the second-to-the-last birth cohort subject to the military draft; were the product of 1950s and 1960s American prosperity, patriotism, and Cold War fears; and were part of the nation's 1960s and 1970s youth culture.

Chapter 4 is the heart of the book. It addresses research Objective 4 by examining the impact that the three social contexts had on the subjects' young adult self-esteems, after controlling for selection bias, context choice predictors, prior self-esteem, and other background variables. Chapter 5 summarizes the study's findings, places them within the broader life course literature, and suggests directions for future research.

Now that we have seen the menu, it is time to sample the fare.

TIMOTHY J. OWENS

# Acknowledgments

This book was several years in the making. Parts of it appeared in articles I wrote for various peer reviewed journals, including the *American Sociological Review*, *Social Psychology Quarterly*, *The Sociological Quarterly*, and *Youth & Society*. I gratefully acknowledge my dissertation advisor at the University of Minnesota, Professor Jeylan Mortimer, for her mentorship, professionalism, and boundless enthusiasm. She has been an inspiration, and my dissertation inspired this book.

I would like to formally acknowledge the direct assistance of the Dean of the School of Liberal Arts at Purdue University, Toby Parcel; my department head, Professor Viktor Gecas; and the selection committee of the Purdue University Center for Behavioral and Social Sciences Research, which awarded me a semester-long fellowship enabling me to complete the writing of this book. Stephanie Young assisted me in innumerable ways, big and small, by doing library work, typing tables, drawing figures, and proofing. Based on her work ethic and intelligence, I am certain she has a very bright future in academe. Heather Rodriguez and Susan Owens also did helpful proofreading just in the nick of time.

Finally, I wish to thank my family. First and foremost, Susan Burns Owens, my wife of over 24 years, has been a great source of help and understanding as I pursued my academic profession. Our three children are also tremendous sources of pride and inspiration for me. Thank you Caitlin, Peter, and Sam. I also thank my parents, Clarice and the late Eugene

Owens for all they have done for me, and my later mother-in-law Betty Burns. Last, I wish to acknowledge my cousin James Victor Mudge, Jr., Jamey to all who knew him, for being a great friend and terrific cousin. He was a class of 1969er, too, who tragically died of lung cancer while I was writing this book.

# Contents

# 1

# Implications of Context Choice for the Early Life Course

## Introduction

Three large and diverse areas of the literature, and their associated sections in this chapter, inform my study of how social contexts may influence self-concept change during the transition to adulthood. The first section examines individual constancy and change over the life course, including psychological, social psychological, and sociological perspectives. The psychological contributions focus on the stage theories of human development found in the works of Erikson (1963, 1982) and Levinson (1978). Attention is also paid to aspects of self-concept formation. The social psychological contributions assess the social contexts as a source of attitude, value, and belief change or stability over time as people move through their lives. Attitude formation over the life course serves as the backdrop for this discussion, with particular attention to Glenn's (1980) aging-stability hypothesis. The sociological contribution to life course research is expressed chiefly through Dannefer's (1984) sociogenic thesis, although the aging-stability hypothesis also has a strong sociological component.

The second section reviews contemporary perspectives on socialization. The sociogenic thesis, aging-stability hypothesis, and House's (1981) model of social structure and personality all emphasize the need to incorporate micro- and macro-social processes in our understanding of human development. In terms of the micro-social processes, attention centers on

1

the contexts and contents of socialization to roles, particularly during adolescence and adulthood.

The final section of the chapter examines issues related to occupational choice. Although the choice of work, military, or college after high school involves a somewhat different set of issues than the choice of a specific occupation, the choice literature is examined because it provides useful theoretical insights that can be applied to my assessment of the movement into social contexts. The occupational choice literature review is divided into three areas: (1) the psychological contributions as expressed by the vocational interest theories of Super (1957, 1984) and Holland (1997); (2) Blau, Gustad, Jessor, Parnes, and Wilcock's (1956) landmark social psychological conceptualization of occupational choice; and (3) the structural contribution illustrated by Granovetter's (1983, 1974) emphasis on weak tie contacts in occupational choice. The occupational choice literature guides the formulation of the context choice scheme presented in chapter 3.


## Perspectives on the Life Course and Human Development

### *Life Stage Viewpoints*

Interest in the stages of life has been a long-standing concern in the history of Western ideas, going back at least to Shakespeare's depiction of the seven stages of the life cycle in *As You Like It*. Briefly, the stage theories of human development tend to hold two central positions. First, it is assumed that one's core identity crystallizes in adolescence and early adulthood. Second, stage theorists assume that human development progresses through well-defined age-specific stages involving core tasks or crises. It is generally accepted that stage progression cannot proceed if earlier tasks or crises have not been completed or have not been successfully resolved. Failure to meet the tasks or crises associated with a particular stage typically results in negative personality outcomes and either regression into earlier stages or stagnation in one's current stage. The three main life cycle stages in the West are childhood (including infancy), adolescence, and adulthood, with various substages within each. Although the temporal ordering of the various tasks or crises characterizing a particular stage are often varied and dependent upon one's culture, a core assumption of stage theorists is that the sequencing of the stages is predetermined (Erikson, 1982, pp. 66–67).

The review of Erikson's and Levinson's stage theories serves as an important foundation from which to examine Glenn's (1980) aging-stability hypothesis and Dannefer's (1984) sociogenic thesis.

*Erikson's Eight Ages of Life*

Erikson (1963, 1982) uses a psychological classification to mark his eight ages of life stemming from psychoanalytic thought (1963, pp. 247–274). Even though he divides the life cycle into eight developmental stages, Erikson refers to infancy, adolescence, and adulthood as the "strategic stages" where the crucial human strengths of hope, fidelity, and care emerge (1982, p. 58).

Each life cycle stage involves a psychosocial crisis involving a particular developmental task appropriate to a given age. Successful resolution of a stage crisis allows progression to higher stages. Those unable to resolve a psychosocial crisis may become stuck and unable to incorporate the exigencies of their "age" in their ego and identity. During the childhood phases of the life cycle one moves essentially with or against biological rhythms. In the later stages, beginning with adolescence and going through the adulthood stages, one is increasingly faced with adapting to roles and situational pressures associated with responsibility for others. But in every stage, the individual is repeatedly faced with a conflict between polar types representing positive development and stage progression or regression and the alternative of cynicism, isolation or despair.

The first three stages are the pre-school years of infancy, early childhood, and the play age. During these stages the child's basic social orientation is the family. In the first two stages the child is chiefly occupied with coming to terms with issues such as trust, autonomy, guilt, and inferiority. *Infancy* is the strategic stage where the child develops basic trust or mistrust of the outside world. "Mothers create a sense of trust in their children by that kind of administration which in its quality combines sensitive care of the baby's individual needs and a firm sense of trustworthiness within the trusted framework of their culture's life style" (Erikson, 1963, p. 249). Infants who do not develop a basic sense of trust tend to exhibit the core pathology of withdrawal.

In the *early childhood* stage, the toddler either begins to develop a sense of autonomy or falls victim to a sense of shame and doubt. The important task here is to develop the ability to freely pick and choose which objects and relationships to hold onto and which to let go. In this stage the child begins to learn discrimination in his or her powers and tastes. An undeveloped sense of autonomy may result in a sense of shame and doubt manifested as compulsive manipulation of self and others in a strive for possessions and control.

In the play age, the child begins to exhibit new social and interpersonal powers (stemming from increased motor and mental abilities). The psychosocial crisis in this stage is between initiative and guilt. A sense of initiative results in "anticipatory rivalry" (Erikson, 1963, p. 256) with older siblings or others in the family environment. However, the child must reconcile the fantasy of being all powerful or being totally destroyed by those with whom he or she competes. Resolution of the antipathies of this stage helps produce a sense of initiative, while failure to do so may result in an inhibited child filled with a sense of guilt.

In the school age and adolescent stages, the young person's social orientation is extended to school and peer groups. The pre-pubescent *school age* child's basic psychosocial crisis is between industry and inferiority (Erikson, 1963, p. 258). Here the child learns the "technical and social rudiments of a work situation" (Erikson, 1982, p. 75) which will enable him or her to be a productive member of society. Industry is where the child "learns to love to learn as well as to play—and to learn most eagerly those techniques which are in line with the *ethos of production*" (Erikson, 1982, p. 75, emphasis in the original). This ethos engenders a sense of cooperation with others in followings plans and scheduled procedures. The antithesis of industry is a sense of inferiority in which the child may be excessively competitive. Instead of feeling competent as in industry, the child develops a sense of incompetence manifested as selfish competition or a retreat from the emerging ethos of production with a concomitant lapse into the core pathology of inertia.

*Adolescence* is the second strategic stage in the life cycle when the post-pubescent child struggles with defining his or her identity. According to Erikson, the "reliability of young adult commitments largely depends on the outcome of the adolescent struggle for identity" (1982, p. 72). The adolescent's important psychosocial task is to selectively affirm or repudiate earlier childhood identifications. Here one seeks a new identity by experimenting with various roles and rebellions, while concern with the perceptions of others increases. The core strength of adolescence is a maturing faith in mentors and leaders, which Erikson calls fidelity (Erikson, 1982, p. 73). The antithesis of identity is identity confusion. If the adolescent's changing self-image and childhood experiences are not reconciled in this stage, identity confusion and the loss of a lasting sense of self may result. An uncrystallized sense of self may have serious ramifications for the life course because it will not only block progression to the next developmental stage, it may also hinder the adolescent from acquiring age-appropriate roles. The acquisition of age-appropriate roles contributes to identity and value formation.

Adulthood is divided into young adulthood and mature adulthood (or simply adulthood). In the *young adult stage*, the post-adolescent faces the psychosocial crisis of intimacy versus isolation. During this stage, one must learn to accommodate oneself to people from different backgrounds and habits and to new milieus, especially upon marriage (Erikson, 1982, pp. 71–72). Central to this important task is the ability to be intimate. Erikson writes (1982, p. 70):

> Young adults emerging from the adolescent search for a sense of identity can be eager and willing to fuse their identities in mutual intimacy and to share them with individuals who, in work, sexuality, and friendship promise to prove complementary. One can often be "in love" or engage in intimacies, but the intimacy now at stake is the capacity to commit oneself to concrete affiliations which may call for significant sacrifices and compromises.

The psychosocial antithesis of intimacy is isolation, or a fear of remaining separate and unrecognized (Erikson, 1982, p. 70). The most serious drawback of isolation is a "regressive and hostile reliving of the identity conflict" of adolescence or a fixation on the crisis of trust versus mistrust engendered in one's relationship with the "maternal person" in stage one (Erikson, 1982, p. 71). Failure to resolve the intimacy versus isolation crisis may result in self-limiting behavior manifested as elitism. Another negative consequence of isolation is the inability to reject or exclude anything, which may have the dire consequence of self-rejection or self-exclusion.

In the *mature adult stage*, one faces the new crisis of generativity versus stagnation. Generativity means being a fully productive and contributing member of society and one's communal group. The generative adult builds and cares for a family, and creates and nurtures new ideas and new products (Erikson, 1982, p. 67). In the process of this productivity and care, a self-generation occurs which helps to further crystallize and define the adult's identity. The antithesis of generativity is stagnation, which is an extension of the isolation and exclusivity described in reference to young adulthood. Stagnation engenders self-absorption and results in the core pathology of rejectivity, or the inability or lack of interest in extending oneself to others or in putting self-interest aside in an attempt to work cooperatively (Erikson, 1982, p. 67).

Finally, in *old age* the culmination of previous stage crises reaches its apex or nadir as the old person faces the psychosocial crisis of integrity versus despair (Erikson, 1982, pp. 64–64). Integrity builds on the successful completion of earlier stages and provides a sense of coherence and wholeness even in the presence of a loss of physical, mental or social powers. Integrity, in short, involves coming to terms with one's past without bitterness or remorse.

Despair, and its behavioral counterpart, disdain, connotes a sense of a life misspent, of time being too short to put oneself on the road to integrity (1963, p. 269). Despair is a loss of hope for a meaningful old age (1982, p. 62).

*Levinson's "Life-Cycle Eras"*

Levinson's (1978) model of the male life-cycle owes much to Erikson's pioneering work but departs sharply from the purely psychological view of the life cycle by incorporating social psychological principles. (Levinson, 1996, shows that his model is appropriate to women as well.) Levinson's model is social psychological because it addresses how the roles and responsibilities of particular life-cycle phases influence one's adaptation to, and movement through, subsequent life stages. Levinson argues that the life cycle is not marked by a "continuous process of development" but is marked instead by "qualitatively different periods in development" (Levinson, 1978, p. 40). He says that from the end of adolescence to the middle forties there is a sequence of periods through which all men must pass. In order to understand the adult life cycle, one must look at the general character of a man's living, not solitary aspects of his life.

Levinson and associates (1978) divide the male life-cycle into four eras with accompanying transition phases. Their "life-cycle eras" and the developmental periods (designated by a "•") are (p. 57):

1. Childhood and adolescence, ages 0–22
2. Early adulthood, ages 17–45
   • Entering the adult world
   • Age 30 transition
   • Settling down

3. Middle adulthood, ages 40–65
   - Entering middle adulthood
   - Age 50 transition
   - Culmination of middle adulthood
4. Late adulthood, ages 60+

The transition phases are (p. 20):
   a. Early childhood transition, ages 0–3
   b. Early adult transition, ages 17–22
   c. Mid-life transition, 40–45
   d. Late adult transition, 60–65

While Erikson focuses on childhood and adolescence, Levinson is concerned mainly with the adult phases of the life cycle. As a result, Levinson merges childhood and adolescence together in his model, while noting that they are distinct stages (Levinson, 1978, pp. 20–21). He refers to them as pre-adulthood: the era in which intensive socialization and psychological and biological maturation combine to provide a sense of self, decreasing dependence on one's family of origin, and sexual identity (Levinson, 1978, pp. 20–21). During the early adult transition period, the pre-adult self is relinquished as the "boy-man" make choices which will enable him to establish "initial membership in the adult world" (Levinson, 1978, p. 21).

*Early adulthood* begins around age 17 or 18 and ends at about age 45. During this era a man typically reaches his intellectual and physical peak, although Levinson believes that one's intellectual peak may also vary with one's career and other factors (Levinson, 1978, p. 22). Early adulthood is distinguished from later eras by "fullness of energy, capability and potential, as well as external pressure. In it, personal drives and societal requirements are powerfully intermeshed, at times reinforcing each other and at times in stark contradiction" (Levinson, 1978, p. 23). As the "novice adult" moves through the early adult era, he is said to become gradually more understanding and responsible (Levinson, 1978, p. 22). But this is also a particularly stressful time of life. Although the early adult usually has not attained his peak earning power, his financial obligations may be the greatest of his life. And when children arrive, his married life also changes and becomes more complicated. By the time the young adult reaches middle adulthood, the previous stresses tend to abate as he attains "senior" status in his kinship group, work group, and community.

*Middle adulthood* extends from about age 40 to 65. Although Levinson tends to idealize this era by implying that nearly everyone who does not fall victim to injury or illness becomes a more compassionate, reasonable, and loving person, his depiction touches upon issues useful for a sociological interpretation of middle-age. During this era the male generally takes stock of himself. His relationship with his spouse, his children (now largely grown or adolescents), and his parents change as family roles change and as others move through their own life-cycles. At work, the middle-age man may still be on the way up, but more than likely he has reached a career plateau. While he may have been driven in earlier adulthood by abundant energy and career or family aspirations, by middle adulthood many of these events have come to pass. In this light, the middle-age man has an objective criterion upon which to compare youthful aspirations with current reality. In short, the middle age man has the opportunity for "self-renewal and creative involvement in one's own and other's lives" (Levinson, 1978, p. 33).

Finally, the era of *late adulthood* lasts from about age 60–85. Middle and late adulthood are not distinguished by a single universal event, like puberty marking the onset of adolescence. (Riley, Johnson, and Foner [1972], to the contrary, show that retirement is a major "marker" in one's passage into late adulthood.) However, late adulthood is, according to Levinson, marked by a recognizable decline or loss in some middle adult powers. During late adulthood, people often experience their first major medical problems such as cancer and heart disease (Levinson, 1978, p. 34). Moreover, Levinson posits that men in the late adult transition period may come to fear that the "youth within him is dying and that the old man—an empty, dry structure devoid of energy, interests or inner resources—will survive for a brief and foolish old age" (Levinson, 1978, p. 35). Within the social structure, the old person "receives less recognition and has less authority and power. His generation is no longer the dominant one" (Levinson, 1978, p. 35). If he cannot let go of his former authority, he may become a "tyrannical ruler—despotic, unwise, unloved and unloving—and his adult offspring may become puerile adults unable to love him or themselves" (Levinson, 1978, p. 35).

*Summary*

Dividing the life course into stages and substages is commonplace (e.g. Settersten, 2003; Levinson, 1978; Riley, Johnson, and Foner, 1972; Sears, 1981; Glenn, 1980). The justification for doing so is typically based on common sense: The 14 year old adolescent, having recently attained puberty, moves in a different social orbit than the 17 or 18 year old adolescent who may be driving a car, working, dating, or leaving high school (e.g., Mortimer, 2003; Simmons and Blyth, 1987). The young adult may be experiencing the social pressures of marrying and starting a family, completing formal schooling, and finding a steady job (Uhlenberg and Mueller, 2003; Hogan, 1981). In contrast, the older adult may be concerned with such things as career leveling or retirement, renegotiating relationships with grown children, or coming to grips with the death of a spouse (Moen, 2003; Settersten, 1999). The underlying assumption guiding the division of the life cycle into various phases is that values, attitudes, and priorities can and do change as people confront the realities of leaving earlier life-stages and entering new ones. While there is little debate over viewing the life course as a series of overlapping phases a good deal of controversy exists over the issue of constancy and change over the life course (e.g. Caspi, 2000; Mortimer and Simmons, 1978; Brim and Kagan, 1980). The next section takes up this issue.

## *Life Course Hypotheses*

This section sets the stage for an understanding of the life course from a sociological standpoint, while accepting, selectively, the contributions made by psychologists. Although some have been known to the academic community for several years, I consider them sufficiently important to frame my current thinking about change over the (early) life course. The first part of this section presents four hypotheses on attitude change through the life course originally outlined by Sears (1981). The second section discusses Glenn's (1980) useful aging-stability hypothesis which informs the substantive analyses presented in chapter 4. Finally, the third part of this section revisits Dannefer's (1984) sociogenic thesis, another useful heuristic for a sociologically-informed view of human development research.

### *Sears' Hypotheses about Attitude Change through Stages of Life*

Sears (1981) identifies four general hypotheses dealing with attitude change over the life course. Although he admits the hypotheses may represent oversimplifications of complex processes (p. 183), they are useful foundations for an examination of the aging-stability hypothesis and the sociogenic thesis. Please keep in mind that Sears presents his four hypotheses with a special eye toward their potential for empirical testing. The hypotheses are: (1) the "lifelong openness" hypothesis, (2) the "life-cycle" hypothesis, (3) the "impressionable years" hypothesis, and (4) the "persistence" hypothesis (Sears, 1981, pp. 183–185).

The *lifelong openness hypothesis* states "age is irrelevant for attitude change" (pp. 183–184). According to this view, people are just as likely to change an attitude at 18 as they are at 35 or 75. People are believed to be malleable and open to change throughout life, even though there may be some initial resistance to change. (For example, this may include resistance to new ideas arising throughout the life course.) Van Maanen (1976, p. 68) and Kanter (1977) show that as a person moves across organizational settings—such as a job promotion or demotion—he or she may be required to relinquish some currently held attitudes, values, and behaviors appropriate in a previous setting and adopt new ones. And this change is not necessarily confined to young or old workers. Mortimer and Simmons (1978), however, cite evidence from the adult socialization literature which suggests that adults have a greater ability to resist socialization pressures than adolescents do (p. 425). In view of the evidence, the lifelong openness hypothesis may need modification; however, it will be shown later that its complete dismissal is not justified, except perhaps in its extreme form. (See Wethington [2002] for a detailed look at the role of small and large "turning points" as a catalyst to change.)

The *life-cycle hypothesis* asserts that "people are particularly susceptible to certain attitudinal positions at certain life stages . . . " (Sears, 1981, p. 184). This is exemplified by the popular notion that youth is a time for liberalism, while middle age is marked by conservatism (Glenn, 1980, p. 619). Sears refutes the life-cycle hypothesis by stressing that most cohort analyses which have tested this hypothesis have "almost invariably found generational rather than life-cycle effects" (p. 188). Cohorts do not respond to age-specific attitudes through the life course, it is the attitudes that a cohort carries into adulthood that largely persist (p. 188). Glenn's (1980, pp. 618–623) research also supports Sears' conclusion. Glenn (1980, p. 619) shows that researchers who link increased conservatism with later life-cycle stages often assumes that a "social aging" process is behind the shift. The argument is that as one moves from adolescence to adulthood, marriage and childbearing responsibilities draw one's attention away from humanitarian ideals and focus it on immediate responsibilities. As people move through the life-cycle their interests tend to shift toward maintaining the status quo, which results in older adults professing more conservative views than do persons in earlier life-stages (Glenn, 1980, pp. 619–620). However, disentangling period, cohort, and generational effects makes empirical testing of the life-cycle hypothesis problematic (see also Glenn, 2003).

Sears may be too quick to dismiss the life-cycle hypothesis, primarily because of the way he frames it. The argument that one may be susceptible to specific attitude positions at different life stages suggests vulnerability and passive attitude shifts. Quite the opposite may happen. Glenn (1980) indicates that a shift in certain political attitudes, for example, may be the consequence of different self-interests at different periods of life. Sears believes that it is pressure to change, along with the amount of ego-involvement a person has toward an attitude, that are the salient features in change. In contrast, Glenn (1980) suggests that different sets of priorities are associated with different life-stages, independent of one's particular self-identity. In this light, people in retirement may shift their previous attitudes about government support of social security or medical benefits and mandated retirement ages. Younger workers, on the other hand, may have a vested interest in forced retirement and reducing taxes stemming from medicare. More will be said about the "life-cycle hypothesis" in the discussion of the aging-stability hypothesis.

The *impressionable years hypothesis* states that people in late adolescence and early adulthood are "unusually vulnerable . . . to changes in any attitudes, given strong enough pressure to change" (Sears, 1981, p. 184). Sears expresses a preference for this view over the preceding two. The literature tends to support the idea that adolescence and early adulthood are important formative stages in the life course (e.g. Mortimer, 2003; Brim, 1966; Elder, 1999; Elder and Rockwell, 1978; Glenn, 1980; Levinson, 1978; and Riley, Johnson, and Foner, 1972). Ryder (1965) believes that people in late adolescence and early adulthood show a greater tendency to change attitudes than people in later adulthood because they are *exposed* to more situations which influence their attitudes—*not* because they are inherently more prone to change. This is perfectly consistent with symbolic interactionism. Here, the key to individual change is the frequency and intensity of acquiring and relinquishing salient roles and identities, as well as traversing various social contexts and networks.

A variation of the impressionable years hypothesis is the "generational effect" hypothesis (Sears, 1981, p. 184). This position is expressed in Mannheim's (1952) sociology of generations. The dominant social and economic conditions experienced by a cohort in its youth may give rise to a prevailing worldview or value structure which sets a particular cohort apart from others. Mannheim (1952) wrote that when society faces a period of rapid change due to war, revolution, or cultural or social upheaval, the events tend to affect various birth cohorts in different ways. The concept of generation is important to Mannheim because it links history and social structure: As each new cohort comes into "fresh contact" with the historical world around it, it has the potential for being changed and stratified in relation to other birth cohorts, through exposure to common events (Mannheim, 1952, pp. 292–320). And, birth cohorts may be differentially affected along such lines as race, class, and residence. Just as Mannheim asserts that overarching historical influences sometimes produce a generational mentality, the generational effect hypothesis states that an entire cohort may be influenced by a "common massive pressure to change on some particular issues" due to exposure to extraordinary events (Sears, 1981, p. 184). Elder's (1999) study of men and women moving into childhood and adolescence during the grim years of the Great Depression is a noteworthy example of research which investigates this hypothesis.

The *persistence hypothesis* is apparent in much life course research, especially research that takes a developmental perspective on stability and change (e.g., Levinson and Erikson). By asserting that preadult attitudes tend to be relatively immune to change over time, this hypothesis posits that age has little effect on attitude change beyond the childhood years (Sears, 1981, p. 184). There is an extensive literature on the process of childhood socialization and the acquisition of values. Gecas (1981, p. 170), for example, writes:

> In the minds of many people, the family is the context most closely associated with the topic of socialization. As an institution, its major function in most societies is the socialization and care of children. The goal of most parents is to develop children into competent, moral, and self-sufficient adults. This is typically undertaken with two objectives or frames of reference in mind: socializing the child for membership in the family group, and socializing the child for membership in the larger society .... This is why the socialization that takes place here is usually the most pervasive and consequential for the individual. It is also the first socialization context that most of us experience— the place where we develop our initial sense of self.

Childhood is thus a time when many people adopt lifelong values (such as religion) and a sense of self.

However, considerable evidence in the sociological literature links changes in adulthood to adult socialization and experience (e.g. Settersten and Owens, 2002; Brim, 1966; Bush and Simmons, 1990; Goffman, 1961; Mortimer and Simmons, 1978; Whyte, 1981; and Young and Willmot,

1962). An extensive overview of the sociological literature on socialization reveals two significant issues (see Settersten and Owens, 2002). The first concerns the amount of continuity and consistency of personality traits over the life course. Most sociological approaches stress the general malleability of the self, and that most definitions of socialization imply the possibility of change (Owens, 2003). The second issue concerns the similarity between socialization processes in childhood and adulthood. Child and adult socialization experiences primarily involve differences in content and context (i.e. social structural properties), rather than processual differences (i.e. learning and cognitive development issues). As we shall see, Dannefer (1984) is particularly critical of life course research which relies too heavily upon age-graded change, without sufficiently attending to the social contexts in which the change is produced. Since the persistence hypothesis is a key element in the aging-stability model, it will be discussed in more detail later.

In summary, Sears' four hypotheses on attitude change are useful for at least two reasons. First, they represent the diversity of opinion which still exists in life course analyses of attitude change. The lifelong openness hypothesis, which stresses the malleability of people's attitudes, forms one pole. The persistence hypothesis, which argues that preadult socialization creates attitudes which are relatively immune to change over time, marks the other pole. Clausen's (1993) research on the role of "planful competence" established in childhood and early adolescence as shaping adults' life course histories of cumulative advantage and disadvantage is illustrative of the persistence hypothesis. Sears (1981) concludes that the lifelong openness hypothesis is inconsistent with most relevant data and has rarely been supported. He considers the "persistence" and the "impressionable years" hypotheses most worthy of continued investigation. His preference for these two perspectives reflects his interests in sociology and psychology. He believes that the lagged effects of childhood and adolescent socialization best account for the pervading "political and social attitudes most important to ordinary people and most consequential for society" (Sears, 1981, p. 191).

The second reason for reviewing Sears' hypotheses is to demonstrate the need to employ a theoretical view of adult development which includes both developmental and environmental influences. Sears believes that environmental influences (i.e., social networks, mass media influences, and historical changes in attitude objects) have more impact on attitude change than do developmental influences (i.e., stage-specific self-interest), but that both should be considered. The aging-stability hypothesis brings these two influences together in a single perspective. I take up that hypothesis next.

*The Aging-Stability Hypothesis*

The aging-stability hypothesis is a theoretical orientation which holds that "attitudes, values, and beliefs tend to stabilize and to become less likely to change as a person grows older" (Glenn, 1980, p. 602). This is an important assertion in life course research, and incorporates aspects of Sears' attitude change hypotheses. In Glenn's formulation of the aging-stability hypothesis, the distinction between attitudes, values, and beliefs is important to note. An attitude is similar to an opinion and is the most malleable of the three cognitions under consideration here (Glenn, 1980, p. 597). He defines an attitude simply as "an evaluation of an object" (p. 597). An object refers to both material things and persons.

Festinger (1957, p. 91) writes that cognition is

> the things a person knows about himself, about his behavior, and about his surroundings . . . . Some of these elements represent knowledge about oneself: what one does, what one feels, what one wants or desires . . . . Other elements of knowledge concern the world in which one lives: what is where, what leads to what, what things are satisfying or painful or inconsequential or important, etc.

According to Glenn (1980, pp. 597–98), value is a "special kind of attitude—one with a highly abstract and general object" (p. 597). Values often are expressed in statements of worth: good or bad, right or wrong, desirable or undesirable (p. 597). The thing which distinguishes a belief or attitude from a value is that a value involves cognition *and* evaluation of a highly abstract object. According to Glenn's use of the term, self-esteem would be a value—the cognition of an abstract object (i.e., the self) and the concomitant evaluation of its worth. For the sake of simplicity, I will use the generic term attitude to refer collectively to attitudes, values, and beliefs, unless indicated otherwise.

Glenn (1980, p. 605) shows that the conceptualization of the inherent changeability of attitudes, values, and beliefs is complex since some attitudes are more stable than some beliefs, some beliefs more stable than some attitudes, and so on (p. 605). He sets forth a "stability-malleability continuum" (p. 605). On the stable end of the continuum are values. Many are initially formed during childhood socialization (see Dube, Felitte, Dong, Giles, and Ando, 2003, and Edwards, Holden, Felitti, and Anda, 2003, for recent examples). These are the so-called "deeply ingrained values concerning religion, the family, marriage," and so on (Glenn, 1980, p. 605). They also include more abstract notions such love, freedom, and democracy (p. 605). The malleability end of the continuum tilts towards beliefs. These include the "nature of changeable and tangible objects" (p. 605), such as one's attitude toward a presidential candidate, especially month-to-month during a presidential election year. The question may not be whether a particular attitude is more or less likely to change; the important question may be whether the *object* of an attitude is more or less prone to change over time. Glenn points out that a person with normal perceptual acuity— regardless of age—will change his or her beliefs about an object, if, over the course of time, the nature of the object itself changes, and providing he or she is interested enough to pay some attention to it (p. 605).

The more abstract a belief, the more stable it is. That is, people don't generally change their beliefs about things they can't readily grasp. A Catholic, for example, may change his or her attitude about Catholicism, and turn away from it, but retain belief in God. Glenn concludes that in general, "values should be relatively stable and beliefs should be relatively changeable. Attitudes should be highly variable, some being almost as changeable as the most changeable beliefs" (Glenn, 1980, p. 605).

In view of this variability, Glenn finds that attitudes are not always appropriate for testing the aging-stability hypothesis. He says instead that, "one must seek measures of attitudes that tend to reflect 'basic values,' and these are generally attitudes with stable or highly abstract objects" (Glenn, 1980, p. 605). Using basic values to test and examine the aging-stability hypothesis, according to Glenn, is important because too much of the social science literature in this area is based on "commonsense . . . and folk wisdom [rather] than on scholarly theory or on systematic examination of relevant evidence" (Glenn, 1980, p. 602). But he credits advances in adult socialization research with successfully challenging the notion that "'basic views' and a 'basic personality structure' crystallize during childhood" (Glenn, 1980, p. 603). The adult socialization literature tends to concur (Lutfey and Mortimer, 2003).

This reasoning reflects the concerns of Mannheim (1952), Ryder (1965), and Eisenstadt (1956), each of whom stresses the importance of post-childhood (i.e. late adolescence and early adulthood) socialization experiences in value formation and change. Furthermore, Ryder and Mannheim are very clear on the role historical and social events play in attitude change. Mannheim says that during a time of "dynamic destabilization," a change in basic values across generation-units may occur which upsets the prevailing social structure and brings new worldviews and institutions into being. The cumulative effect of such an occurrence may result in the creation of a so-called "generational mentality" (Mannheim, 1952, p. 298) in the cohorts who bear the brunt of the historical events.

A compelling aspect of the aging-stability hypothesis from a sociological standpoint is the idea that it is the spacing of significant life events—and *not* the inherent unchangeability of people in the later years of life—which accounts for the crystallization of their values in young adulthood (Glenn, 1980, p. 603; see also Settersten, 2003, and von Eye, Kreppner, Spiel, and We-Sels, 1995). While some evidence suggests that physiological changes in aging may account for the supposed rigidity of attitudes and behavior in older persons (e.g. Carlsson and Karlsson, 1970), or that a decline in intellectual capacity over the life-span may make it difficult to process new information and experiences (Horn and Donaldson, 1980, p. 468–474), Glenn and others argue that sociological explanations are also important (see especially Riley and Riley, 1999). The logic of the argument runs as follows:

> Major life events bring about changes in values.
> Most major life events are concentrated in late adolescence and early adulthood.
> Therefore, most value change occurs in late adolescence and early adulthood.

This syllogism means that if one compares the degree of attitude change in young adults with the degree of attitude change in older adults, young adults will show more attitude change.

In this context one should view attitude, value, and belief changes within a more encompassing framework. One should look to the historical conditions, social structure, and life events which influence people's values, attitudes, and beliefs (Staudinger and Pasupathi, 2000; Riley and Riley, 1999). House (1981) makes a similar plea in his model of social structure and personality—particularly as applied to what he calls its "components" and "proximity" principles. Briefly, the "components' principle calls for an explication of the dimensions and components of a social phenomenon or social organization thought to influence the individual, as well as an explication of the organization's location in the larger social structure (p. 540). The "proximity" principle links the macro-structures identified in the "components" principle with the micro-structures which have a more direct impact on personality (see also Staudinger and Pasupathi, 2000). For example, House argues that in order to understand individual constancy and change, one must take into account both societal level phenomena as well as how the components of the social structure mediate those influences. House (1981) summarizes his position this way:

> a major theoretical task in the study of social structure and personality is to trace how macro-social structures and processes affect increasingly smaller social structures (for example, formal organizations) and ultimately those micro-social phenomena that directly impinge on the individual (p. 541).

Chapters 3 and 4 follow the "components" and "proximity" principles by examining the social and psychological factors which influence entrance into the social contexts of work, military or college (chapter 3) and also by examining how the social contexts influence personality development (chapter 4).

The aging-stability hypothesis can also be interpreted in terms of the cumulative effects of experiences which produce change (see O'Rand's, 1996, discussion of cumulative disadvantage theory). Glenn (1981, p. 604) notes that "existing attitudes are the products of accumulated experience and . . . their resistance to change varies directly with the amount of experience that has produced and reinforced them." In this sense, if attitudes are formed and reinforced by external stimuli, repeated consistent stimuli (either pro or con) will tend to diminish the likelihood of a change. As a result, the longer a person holds an attitude, the less likely the attitude is to change. Thus, as a person ages, his or her attitudes tend to be reinforced many times over. And if a person's attitudes are constantly reinforced, then it becomes increasingly remote that exposure to a few contrary attitudes will result in attitude change. This happens because disparate attitudes must compete against the hundreds of incidences which confirmed the entrenched attitudes. A young person has had far fewer consolidated reinforcing experiences so his or her attitudes are probably not as ingrained. This fosters greater susceptibility to the socializing efforts of teachers, role models or significant others. The role model, in turn, will add to his or her reinforcing experiences by interacting with the socializee. More is said about socialization later.

### The Sociogenic Thesis

Despite the proliferation of empirical and theoretical research on the life course over the past three decades (e.g., Mortimer, 2003; Mortimer and Shanahan, 2003; Bengstson, Biblarz, and Roberts, 2002; Sampson and Laub, 1996; Simmons and Blyth, 1987; Elder, 1975, 1979, 1999), no overarching theoretical orientation for life course research has yet emerged. Indeed, Elder, Johnson, and Crosnoe (2003) prefer to refer to life course theory as a theoretical orientation. They write:

> [W]e view the life course as a *theoretical orientation*, one with particular relevance to scholarship on human development and aging, and we use the term 'theory' with this particular meaning.... [T]heoretical orientations establish a common field of inquiry by providing a framework for descriptive and explanatory research (emphasis in the original; Elder, Johnson, and Crosnoe, 2003, p. 4)

They define the life course "as consisting of age-graded patterns that are embedded in social institutions and history.... [It] is grounded in a contextualist perspective and emphasizes the implications of social pathways in historical time and place for human development and aging" (Elder, Johnson, and Crosnoe, 2003. p. 4).

In a major critique of the life course and human development literatures two decades ago, Dannefer (1984) pled for an interdisciplinary perspective on adult development and the life course that incorporates sociological principles. In that article he advocated a developmental perspective that he calls the sociogenic thesis. The sociogenic thesis holds that: (1) the human organism develops in relation to its environment, so that what may appear developmentally normal, is actually socially produced; (2) social environments are structurally complex and diverse and individuals both mold and are molded by their environments; and (3) social knowledge and human intentionality are mediating factors in human development (Dannefer, pp. 106–107). Ontogenetic, or age-related, interpretations of adult development are flatly rejected by Dannefer.

Dannefer (1984) observed that many models of adult development contain an 'ontogenetic fallacy' because they "... surrender ... socially produced age-related patterns to the domain of 'normal development'" (p. 101). He believed that the ontogenetic fallacy stemmed from an intra-disciplinary perspective on the life course which did not include a well developed sociological component. Rather than looking at the profound influence that social contexts play in development over the life course, much of the literature he reviewed followed an ontogenetic (age-related) model consisting of three key assumptions: (1) the individual is treated as a self-contained entity, (2) the "profoundly interactive nature of self-society relations" goes largely unexplored, and (3) the complexity and variability of social environments is largely ignored (Dannefer, 1984, p. 100). In many of these studies, individuals were placed on psychological timetables characterized by age-related psychological dispositions and developmental tasks appropriate to various stages of the life course. In short, Dannefer believed that an overemphasis on age-graded change existed which did not adequately account for the role that social context and life events play in adult development.

To illustrate his point, Dannefer drew on three major developmental approaches in psychology to which sociologists attach varying weight or importance: stage theory (e.g., Erikson, 1963; Levinson, et al., 1978), life-span developmental psychology (e.g., Baltes, 1979), and dialectical life-span psychology (e.g., Gergen, 1980, p. 101). His general criticism of these three developmental approaches is that they succumb too often to the "ontogenetic fallacy," regardless of claims or intentions to the contrary. Furthermore, each perspective has a unique set of theoretical or methodological problems that render them either useless or weak theories.

According to Dannefer, *stage theory* is weakened on theoretical grounds because it does not fully take into account the importance of life events and social contexts when interpreting developmental patterns at various life stages. Further, stage theories may actually encourage the researcher to use *post hoc* judgments in an attempt to fit the data to the stages set up *a priori* (Dannefer, 1984, pp. 103–104). On methodological grounds, the theory is criticized because it implies a simple, universal relationship with age as the independent variable and developmental sequence as the dependent variable. However, age is confounded with specific historical conditions, as well as social and cultural patterns (e.g., career and family cycles).

However, Dannefer's critique seems rather harsh, particularly as applied to Levinson's version of stage theory. As discussed earlier, Levinson (1978) posits a model of the male life-cycle (and later the female life-cycle [Levinson, 1996]) which combines psychological and sociological themes. The theoretical perspective that Levinson applies to his work is the "individual life structure" which he defines as the "underlying pattern or design of a person's life at a given time" (Levinson, 1978, p. 41). He identifies three interrelated components of the life structure: the sociocultural world in which an individual acts and is acted upon (e.g., his social contexts, cultural milieu, and community), the individual's self (e.g., his attitudes, moral values, and skills and talents), and the individual's actual participation in the world (e.g., the give and take which the roles he assumes demand) (p. 42). Levinson writes:

> I am not talking about stages in ego development or occupational development or development in any single aspect of living. I am talking about periods in the evolution of the individual life structure. The periods, and the eras of which they are part, constitute a basic source of order in the life cycle. The order exists at an underlying level. At the more immediate day-to-day level of concrete action, events and experiences, our lives are often rapidly changing and fragmented (p. 41).

He points out that his theory is neither conformist nor deterministic.

> A valid theory of development is not a mold or blueprint specifying a single 'normal' course that everyone must follow. Its function, instead, is to indicate the developmental tasks that everyone must work on in successive periods, and the infinitely varied forms that such work can take in different individuals under different conditions (Levinson, 1978, p. 41).

The *life-span developmental psychology perspective*, according to Dannefer (p. 104), has many prominent sociologists as contributors. This perspective holds that "development or change is an indeterminate but continuous process involving multiple causes and multiple outcomes" (Dannefer, 1984, p. 104). By seemingly downplaying an ontogenetic view of adult development (a view which treats the individual as a self-contained entity), life-span theory attempts to incorporate environmental factors and history while applying quantitative methods in an effort to separate age effects from cohort effects (p. 104). Life-span development psychology is criticized because "while cohort differences are treated as deriving from historical variations, differences within a cohort are still implicitly treated as ontogenetic or maturational" (p. 104) Further, intracohort diversity is too often treated as error variance, which makes the linkage between social factors and human development problematic (p. 104).

Again, however, Dannefer may have been presenting a oversimplified picture. In a sociological work that falls within this rubric, Elder and Rockwell (1979) found evidence to support the notion that exposure to historical events within different family economic contexts has important implications for both adolescent feelings of well-being and efficacy, as well as mid-life health. They compared the life patterns and adult health of two cohorts of men who had very different experiences during the Great Depression: those who encountered the Depression as young children (the Berkeley, California birth cohort of 1928–1929) and those who encountered it as adolescents (the Oakland, California birth cohort of 1920–1921). Further, they emphasize intracohort variation in economic context and resources, contrary to Dannefer's view of life-span developmental psychology.

Elder and Rockwell (1979) found that the younger cohort was more vulnerable to "family strains and disruptions in the Depression and were exposed to a longer phase of economic hardship and its persistence up to departure from home" after World War II (p. 295). As a result, the men in the younger cohort who were raised in economically deprived families (families which lost 34 percent or more of their pre-Depression family income) had more feelings of psychological inadequacy in adolescence (e.g., lower self-esteem, less well-developed vocational interests, and poorer school performance; and more indecision and passivity) than those from their cohort who were reared in nondeprived families. However, by age 40, men from deprived families in general improved. The older cohort, regardless of their family's economic rating, did not suffer these problems in adolescence and adulthood (pp. 298–299).

The improvement in the men's lives, in the younger cohort, was explained by the maturing influences of work, military, and marriage (Elder and Rockwell, 1979, pp. 298–299). The healthiest men from the younger cohort combined a supportive and nondeprived family background with rewarding adult work achievements. The unhealthiest men tended to be those from the younger cohort who were raised in economically deprived families and were "relatively unsuccessful in their work life" (Elder and Rockwell, 1979, p. 299).

Finally, *dialectical life-span psychology* rejects stage theory (Alexander, 1982, p. 519) and ontogenetic approaches to adult development (Dannefer, 1984, p. 106). Instead, it posits "luck," "autonomy," and "life crises" as the driving forces behind adult development and the life course (see Dannefer, 1984, p. 106; Alexander, 1982, p. 519). In a review of Gergen (1980), Dannefer points out that this dialectical approach "implies that predictions about developmental patterns cannot be made and should not be attempted" (p. 106). Needless to say, dialectical life-span psychology has had little if any impact on sociological work and is roundly dismissed by Dannefer because it ignores the "pervasive impact that social structure [has] as an organizer of development," even though it does attend to historical variation (p. 106).

While Dannefer (1984) sometimes presented rather oversimplified views, he made the valid point that too little attention was generally being paid to the social factors that have consequences for change over the life course (see also Featherman and Lerner, 1985). As we have seen, the aging-stability hypothesis attempts to redress this imbalance. But neither is sufficient. Instead, recognizing life course research as a theoretical orientation drawing from a variety of fields and subfields is instructive. Some 20 years after Dannefer's call for better theoretical clarity, and after Glenn's postulations of the aging-stability hypothesis, Elder, Johnson, and Crosnoe (2003) formulated a useful codification of life course theory that both complements and extends Dannefer's, Glenn's, and Sears' thinking. I turn briefly to that now.

*Life Course Theory Paradigmatic Principles*

Up to this point I have reviewed a number of life course-oriented hypotheses, postulates, and theses. All are useful in helping to frame my study of how post-high school social contexts influence self-esteem development in the transition to adulthood. However, the picture remains incomplete. Elder, Johnson, and Crosnoe's (pp. 11–13, 2003) depiction of five basic life course paradigmatic principles helps flesh out the theoretical underpinnings of my life course research. Their principles are:

1. "*The Principle of Life-Span Development*" (Elder, Johnson, and Crosnoe, 2003, p. 11, emphasis in the original). Here, human development and aging are seen as lifelong processes best studied by examining lives over long periods of time, not just until age 18. By taking a longer view of development, we are better able to understand, among other things, how "patterns of late-life adaptation and aging are generally linked to the formative years of life course development" (p. 11). In addition, taking the long view, coupled with longitudinal analytic techniques, affords the opportunity to "increase the potential interplay of social change with individual development" (p. 11). Although my present research only spans adolescence and early adulthood, it recognizes that change is possible throughout the life course, especially when social and historical context are taken into account. This principle is thus a more nuanced and tempered variation of the lifelong openness hypothesis (Sears, 1981) discussed earlier.

2. "*The Principle of Agency*" (Elder, Johnson, and Crosnoe, 2003, p. 11, emphasis in the original). In line with fundamental symbolic interactionist theory, this principle recognizes that individuals play an active role in constructing their own life course, and are not simply passive beings acted upon by larger social forces and structures outside their control (cf. Wrong, 1961). Nevertheless, while individuals can have an active role in shaping the course and direction their lives take, as chapter 3 points out later, a complete picture requires acknowledging that

decisions and actions take place within the opportunities and constraints afforded by social and historical circumstances.

3. *"The Principle of Time and Place"* stresses the importance of recognizing that individuals and their life courses are *"embedded and shaped by the historical times and places they experience over their lifetimes"* (Elder, Johnson, and Crosnoe, 2003, p. 11, emphasis in the original). This is a central aspect of the research I present in this book. The class of 1969 came of age and entered a post-adolescent world of protest and social upheaval unprecedented in American history, except perhaps the Civil War-era. Moreover, as the latter part of chapter 3 will show, the high school class of 1969 (most of whom are the birth cohort of 1951) are the last group of American males at risk for heavy combat in Vietnam, are the second-to-last birth cohort subject to the draft, and for the veterans at least, served in a military in mission, leadership, and moral chaos (Krepinevich, 1986; Gabriel and Savage, 1979). However, as grim as the preceding may have been, they also left high school when unskilled and semi-skilled jobs were plentiful, inflation was low, abundant fun and travel options were available, higher education was within reach of most young people (whether at a local community college or state-run four year institution).

4. *"The Principle of Timing"* recognizes that *"developmental antecedents and consequences of life transitions, events, and behavioral patterns vary according to their timing in a person's life"* (Elder, Johnson, and Crosnoe, 2003, p. 12, emphasis in the original). To life course researchers this may seems obvious; nevertheless, it is also a central tenet that informs most research on the life course, and certainly the present research as well. This principle adds substantially to the life course hypotheses discussed earlier in reference especially to Sears, by combining sensitivity to age while escaping the ontogenetic fallacy discussed earlier by Dannefer. Although my research, especially on the military context, is quite sensitive to the timing principle, the data's focus on one cohort's early adult experiences negates any real possibility of exploring this principle in my present research. It would, for example, be quite interesting to assess the developmental consequences of military service right out of high school (a very common event for working class males) and service delayed until the early to mid-20s. However, this principle does have some bearing on the work context group. As I define and discuss later, they are composed of boys who, after leaving high school, spent most of the next five years in the full-time labor force and not in college or the military. It is thus possible to assess some of the developmental consequences of entering the adult world of work right after high school. The issue of timing of entry into the full-time labor force, especially in comparison to the other social contexts (military and full-time college), is addressed extensively in chapter 4.

5. *"The Principle of Linked Lives"* (Elder, Johnson, and Crosnoe, 2003, p. 13, emphasis in the original) recognizes, as John Donne did some 500 years ago, that no man is an island. People live their lives interdependent with those of others'

in a network of shared relationships framed by socio-historical influences. The intergenerational transmission of values (e.g., Kohn and Schooler, 1983) fits here, as does Elder's work on children of the Great Depression (Elder, 1999) and Conger and Elder's work on families caught up in the farm crisis of the 1980's (e.g., Conger and Elder, 1994). While this is a key principle, most of the research I present here is not relational, except some of the parental influence material presented in chapter 3.

*Summary and Conclusion*

This section reflects the controversies and debates present in the inter-disciplinary life course literature, particularly surrounding the constancy and change of attitudes. A prevailing theme which crops up time and again is the importance of late adolescence and early adulthood for attitude, value, and belief formation. Sociologists are increasingly emphasizing the importance of understanding constancy and change within a comprehensive life course context with appropriate theoretical models and orientations. The socialization literature presents another alternative to ontogenetic thinking. A review of this literature is presented in the next section.

## Socialization to Roles

The socialization literature is important to consider in an assessment of change and stability through the life course. A guiding assumption of the aging-stability hypothesis, the sociogenic thesis, and the life course principles discussed above is that the social environment has an important influence on human development. This environmental influence is believed to be importantly mediated through a process of socialization to roles.

### *Contexts and Contents*

*Overview*

Regardless of age, socialization is a process that must be viewed from the vantage points of the group and the individual (Mortimer and Simmons, 1978). Socialization on the group level is the mechanism through which "new members learn the values, norms, knowledge, beliefs, and ... interpersonal and other skills that facilitate role performance and further group goals" (Mortimer and Simmons, 1978, p. 422). For the individual, socialization is a learning process that facilitates his or her successful participation in social life. Like Simmel (1950), Park and Burgess (1921), and Parsons (Bourricaud, 1981), most contemporary sociologists agree that the object of socialization is not necessarily to subdue the unique and idiosyncratic personality and behavior of individuals. Instead, because all structured social interactions (e.g. groups) rely on at least minimally stable and predictable behaviors, the object of socialization from the group perspective is largely confined to social behavior, role enactment, or both (Mortimer and Simmons, 1978). Social behavior in this context means overt behavior which has consequences for social interaction; role enactment, on the other hand, means carrying out the rights and obligations of a role.[1]

Because of the numerous positions one may hold in societies marked by high social mobility, role is the key to understanding behavior in groups and society. Many others arrive at a similar conclusion (Brim, 1966; Riley, Johnson, and Foner, 1972; Mortimer and Simmons, 1978; Van Maanen, 1976; Thornton and Nardi, 1975; and Merton, 1976). Heiss (1990) shows that conceptual clarity is enhanced when role (i.e., social role) is distinguished from social behavior. "The ultimate dependent variable in social-psychological theory is social behavior and if roles refer to actual behavior there would be little for roles to explain" (p. 95). Thus, if the aim of social psychology is to explain behavior both in group settings and as a consequence of involvement in groups (i.e. social behavior), then equating social behavior with social roles is to logically confuse the thing being explained (i.e. social behavior) with the thing used to explain it (i.e. social roles). In Heiss' view, "role is a set of expectations in the sense that it is what one *should* do" (p. 95). And the way in which these role expectations are impressed upon a person is through socialization. For example, one may speak of socialization to old age (e.g. Rosow, 1974; Riley, Johnson, and Foner, 1972), to work (Van Maanen, 1976), to gender (Miller and Garrison, 1982), to the military (Moskos, 1976, 1970; Stouffer, 1949, v. 1), and on and on. One may also address the context of socialization (e.g. Lutfey and Mortimer, 2003; Heiss, 1990; Gecas, 1981; Brim, 1966), the content of socialization (e.g. Eder and Nenga, 2003; Van Maanen, 1976; Brim, 1966), and the individual's response to socialization (e.g. Mortimer and Simmons, 1978; Rosow, 1974). Although most socialization research has tended to take a unidimensional viewpoint, where the only interest is how some agent or context of socialization affects a socializee, two-way analyses have started to emerge in the literature (Corsaro and Eder, 1995).

Heiss (1990, p. 104) identifies three orientations in the sociological literature which concern the context in which socialization to roles takes place. By socialization context, he means the age at which socialization occurs, the period of time between learning a role and performing it, and the nature of the social setting in which learning occurs (p. 104–105). Thus, three socialization contexts in which role learning occurs are:

1. The particular life-stage.
   *Example*: childhood, adolescence, or adulthood;
2. A time element consisting of socialization to a role currently held, about to be held, or to one which may only be held in the future (i.e. anticipatory socialization).
   *Example*: learning to be a parent after one's baby is born, in anticipation of having a baby in the near future, or by playing mother or father as a child;
3. Formal versus informal settings.
   *Example*: socialization into the military versus socialization in a friendship group.

Mortimer and Simmons (1978, p. 423) note that the contents of socialization may focus on biological drives; the development of overarching values and self-image; or the teaching of specific norms, behaviors, and skills.

*Socialization in Childhood.* The content of early childhood socialization may include toilet training or respect for parental authority. It is most often carried out in informal settings such as the family or extended kinship group. When a child plays house, for example, he or she may deal with a content of socialization (imitating parents) and a context of socialization (projecting his or her anticipated adult role through time) (Heiss, 1990, p. 105). Other important socialization contexts for children, besides the family, are the school, childhood peer group, and increasingly television (Gecas, 2000).

*Socialization in Adolescence.* In adolescence, the content of socialization emphasizes enduring value structures and self-esteem (Mortimer and Simmons, 1978). The setting of adolescent socialization is often the formal organization of a school; but the adolescent may also begin to be involved in work and informal voluntary organizations such as social clubs (Lutfey and Mortimer, 2003). However, the informal adolescent peer group is especially important during this life-stage (Gecas, 2000). Peer groups allow adolescents to exercise independence from adult control, to develop contra-values, and to express behavior that is disapproved by adults. The adolescent (and child) begins to develop and validate the self, learns how to present the self through role-taking and impression management (Goffman, 1961), and acquires knowledge of topics that may have been avoided by adults during childhood socialization, such as sexuality (Gecas, 2000).

Elder and Rockwell (1979, p. 260) point out that engaging in paid employment during adolescence may have the unintentional outcome of providing the opportunity for "reality testing," which confronts the adolescent with his or her real employment potential, actual level of skill, and prospects for future employment. For Elder and Rockwell, adolescent employment and reality testing further the crystallization of occupational values and commitment, which may foster stable career lines at an early age (p. 260). Mortimer (2003) particularly confirms this, especially teenage work among lower socioeconomic youths. The concept of reality testing certainly implies the socialization to work roles and the enactment of these roles within an organizational setting. Yet this link between socialization to work roles may also raise (or lower) an adolescent's self-esteem. This idea can be inferred from the principle of reflected appraisals (see Rosenberg, 1979). Reflected appraisal is the process by which individuals see themselves through others' eyes and form some kind of corresponding judgment about their relative worth or worthlessness. Applied to the work setting and socialization to occupational roles, it seems reasonable to conclude that when an adolescent learns his or her work role well and is recognized for it by others (especially peers), his or her self-esteem should rise accordingly. This seems especially true given the importance society places on work. And even though the adolescent may make naive mistakes at work, the eventual mastering of occupational roles (regardless of how small) should be a source of both satisfaction and the beginning of adult autonomy (e.g. Owens, Mortimer and Finch, 1996).

*Socialization in Adulthood.* Outside of the family setting, involvement in a work context is the dominant activity for most adult men and women in contemporary American society. As Gecas points out, the primary goal of work organizations is not socialization, but the delivery of goods and services (1981, p. 187). Still, socialization in work settings is very important for adults. Van Maanen (1976, p. 67) notes that 90 percent of the American labor force is employed in an organizational setting. That is, nine out of ten American workers are employed in an organization which has some kind of hierarchy of positions and differentiation of labor. Van Maanen presents an interesting connection between organizational structure and role socialization which is left largely untouched in the literature I have reviewed. According to him, work organizations are essentially conical, with three dimensions linking the individual to the organization (Van Maanen, 1976, p. 78). The vertical dimension connotes an individual's rank or level within the organization. The radial dimension involves an "assessment of one's centrality within an organization" (p. 78); that is, the importance of a person's role to the functioning and success of the organization. The circumferential dimension concerns the individual's functional responsibilities within the organization.

When a worker traverses a particular organizational dimension, in any direction, he or she also makes a boundary passage. Van Maanen (1976) believes that a boundary passage makes the individual more vulnerable to organizational socialization efforts. As a person moves across organizational settings—and by extension, organizational boundaries—he or she may be pressured to relinquish certain attitudes, values, and behaviors that were appropriate in the previous setting (Van Maanen, 1976, p. 67). Kanter (1977) makes a parallel observation. When secretaries in a corporation move "upward" on an organization's vertical dimension—i.e., are promoted from a secretarial pool to an executive secretary position—they are often seen by the secretaries they are leaving as 'disloyal' to them (Kanter, 1977, p. 151). Once situated in their new position they may be socialized there to avoid friendships with their previous co-workers. By doing so, they will appear to be loyal to the interests of other executive secretaries, and especially the executive for whom they now work exclusively. In this sense, the former office secretary leaves the security of her 'secretaries' primary group, where there is a considerable amount of social and emotional support, and enters the secondary group associated with the executive secretary status. (Secondary group in this case means a group formed primarily on the basis of intersecting roles and responsibilities which de-emphasize the uniqueness of the individual.) When she enters the secondary group of the executive secretaries, she also encounters several socialization contents which range from learning the protocol of making luncheon appointments for her boss or herself, to attaching her corporate identity to her new boss' corporate status (Kanter, 1977).

*Conclusion*

This section has focused on the contents and contexts of socialization, with appropriate attention paid to life-stage variations. We now turn to a consideration of the process by which the individual selects one of the most significant socialization contexts to be encountered in adult life: the topic of occupational choice.

## The Process of Occupational Choice

### *Overview*

The occupational choice literature can be grouped into three areas: psychological, social psychological, and structural. In general, the psychological literature is informed by the same sorts of ontogenetic (or age-normed) assumptions that Dannefer noted in reference to the stage theories prominent in psychology. The psychological literature examines stages of decision-making, developmental stages, and rational actor models which allege that people attempt to maximize their occupational rewards (and minimize costs) by achieving the closest possible fit between their personalities and the work environment. I will focus my attention on two dominant contributors to this area—Holland and Super—each of whom brings a unique and influential perspective to bear on occupational choice from a psychological perspective.

The social psychological literature on occupational choice typically seeks, as its name implies, to provide a unified view of occupational choice and labor force entry within a framework that incorporates both psychological and sociological factors. On the one hand, the psychological factors that account for occupational choice include personal preferences, attitudes, aptitudes, and personality types; on the other hand, these psychological factors are placed within the context of broader educational and labor market opportunities, industrial growth rates, economic conditions, employment policies, the historical milieu, demographic conditions, high school tracking, and so on (see Blau et al., 1956; Rosenbaum, 1976). The social psychological orientation to occupational choice is illustrated by examining the work of Blau and associates (1956).

At the structural level of inquiry, Granovetter's (1973, 1974) social network approach is exemplary. Before reviewing his work, however, a basic overview of the structuralist position is presented to place

Granovetter's work in its broader theoretical tradition. Representing a clear departure from the earlier perspectives we will consider, in structural analyses, individuals—and their respective attitudes, aptitudes, and intentions—often pale, if not disappear, as attention focuses on groups and social organizations and the relationships that exist within and between them. In an acerbic structuralist critique of American sociology, Mayhew (1980) argues that sociology *is* the study of social networks, but that

> Most American sociologists do not study sociology in the structuralist sense of the term . . . . Rather, they merely assume the existence of social structures in order to study their impact on *individuals*, that is, in order to study *social psychology* (the study of the behavior and experience of individuals in social stimulus situations . . . ). In other words, most American sociologists adopt the *individualist* perspective in that the individual is their unit of analysis and so-called 'human behavior' (in both its subjective and objective aspects) is the individual level phenomena they seek to explain or interpret [emphases in the original] (p. 339).

Marx (1964, p. 176) summarizes the structuralist position well in a passage quoted by Mayhew (1980, p. 338): 'Society does not consist of individuals, but expresses the sum of interrelations in which individuals stand with respect to one another.' Thus, in the "*structuralist* conception of social life," Mayhew (p. 338) argues, "sociologists are studying a communication network mapped on some human population. That network, the interaction which proceeds through it, and the social structures which emerge in it are the subject matter of sociology."

Each of these perspectives on occupational choice will now be examined.

## Psychological Contributions to Occupational Choice

### Super's Theory of Vocational Choice

Super (1957, 1970, 1984) and Super, Stariskevsky, Matlin, and Jordaan (1963) made major contributions to the psychological literature on occupational choice by wedding a rational choice model to the actualization of the self-concept. Super's theory of vocational choice hinges on the idea that people tend to choose jobs that implement their self-concepts (1957, 1984, p. 208). Super (1957) and Super and associates (1963) extend his theory by assessing movement through a series of five stages, with each stage characterized by specific vocational developmental tasks. In the theory, vocational development becomes synonymous with the development of the self-concept, with "the final selection of a vocation reflecting the thoroughness with which [the individual] has implemented his self concept into the world of work" (Burns, 1979, p. 260).

Super (1957, p. 71ff) postulates five major stages of career development[2], with specific substages within each. Key to each stage is a vision of the rational individual striving to arrive at a vocationally mature relationship between himself and the world of work. But as Super reminds us (1957, p. 184), vocational choice in its broadest sense is not an event, but a process. (All of the authors considered in this section agree with Super on this point.) The process begins in childhood and continues into post-retirement. Before proceeding, it must be noted that the ages associated with each stage are only rough estimates, just as the age boundaries corresponding to Erikson's and Levinson's stage theories are also general guides. The stages and substages are (Super, 1957, pp. 71ff)[3]:

> *Stage 1*: Growth (birth-14); substages: fantasy (birth-14), interest (11–12), and capacity (13–14).
> *Stage 2*: Exploration (age 15–24): substages: tentative (15–17), transition (18–21), trial (22–24).
> *Stage 3*: Establishment (25–44); substages: trial (25–30), stabilization (31–44).
> *Stage 4*: Maintenance (45–64); substages: none.
> *Stage 5*: Decline (65 and on); substages: deceleration (65–70), retirement (71 on).

The accomplishment of the various developmental tasks appropriate to each stage indicates the degree of *vocational maturity* the individual possesses as well as the individual's location along a vocational development continuum (Super, 1957, p. 185). Vocational maturity is a key concept in Super's research because

> it may be described not only in terms of the gross units of behavior which constitute the life stages, but also in terms of much smaller and more refined units of behavior manifested in coping with the developmental tasks of a given life stage. It is this latter definition which is most helpful in considering a given individual who functions at a certain life stage (1957, p. 186).

Vocational maturity, then, is a point along a continuum of vocational development (Super, 1957, p. 187). It is indicated by the person's awareness of his or her actual interests, skills, and abilities, and by knowledge of what the prospective, chosen, or actual occupation really entails (Super,

1957, p. 129–130). Thus, the specific indicators of vocational maturity will vary according to one's particular life stage. Super (1957, p. 187) has identified the following five characteristics of vocational maturity which pertain especially to adolescence:

1. Increasing orientation to vocational choice,
2. Increasing amounts of vocational information and more comprehensive and detailed planning (i.e., planfulness),
3. Increasing consistency of vocational preferences,
4. the crystallization of traits relevant to vocational choices (i.e., skill acquisition) and
5. Increasing wisdom of vocational preferences (i.e., greater correspondence between the self and the vocational choice).

Osipow (1968, p. 124) has noted that Super's five life stages connote a gradual development of vocational concerns:

> starting in late childhood in tentative probes and questions, becoming stronger stirrings in early adolescence as recognition of the importance of these decisions grows, and finally leading to educational, and sometimes preliminary vocational, decisions. These decisions, in turn, are evaluated and either are modified or become crystallized, and lead to the mature stage of elaboration and embellishment of vocational behaviors.

Super (1984, p. 200–203) stresses that while there are specific tasks and decisions to accomplish at each life stage, there is a constant cycling and recycling of five vocational development tasks:

1. *Decline* in desire or ability to engage in some important previous activities (i.e., having less time for hobbies as one moves from adolescence into early adulthood),
2. *Maintenance* of one's chosen occupation,
3. *Establishment* of one's occupational role,
4. *Exploration* of the boundaries of one's occupational role, and
5. *Growth* in one's vocational maturity (if the challenges presented in the exploration stage are faced).

In terms of beginning to define viable and satisfying work roles (and eventually occupations), we must remember that adolescence is a period of major inward and forward exploration. Super says that adolescence literally means growing up and becoming an adult (1957, p. 80). On the one hand, adolescence is viewed sociologically as the process of moving from the subculture of teen-agers to the subculture of adults (Parsons, 1942). On the other, adolescence is viewed psychologically "as the process of finding out what constitutes adult behavior, and it is the process of trying out various modes of adult behavior and of ascertaining which of these [modes of behavior] are both congenial to one's self and acceptable to one's associates" (Super, 1957, p. 80). In this interaction between the psychological and the social realms, the youth both brings his or her conception of self to the workplace and has it shaped and molded therein (Super, 1957).

*Holland's Theory of Vocational Personalities and Work Environments*

While complementary to Super's theory of career development and vocational maturity, Holland's (1997) theory is still quite distinct. Instead of viewing career development as the implementation of one's self-concept through a series of life stages, Holland believes that the choice of vocation is an expression of global personality type (1997, pp. 7–8). Holland centers his attention on defining six personality types which he then matches to six equivalent work environments which he believes give expression to them. When a person's vocational personality is congruent with his or her actual work environment, then the person's personality is said to fit the particular work environment. However, if one's personality is at variance with one's work environment, then the person will probably experience some adjustment and promotion problems in the workplace. Holland believes his model enhances understanding job satisfaction, achievement, and the likelihood of a job change, all of which are a function of the job-personality fit.

According to Holland, vocational interests are actually an important window to, and aspect of, an individual's personality. "If vocational interests are an expression of personality then it follows that interest inventories are personality inventories" (1997, p. 8). And if this is true, then it is reasonable for Holland to conclude that people choose vocations that give expression to their personality. He suggests that people seek work environments that are conducive to their personalities on two levels (pp. 7–14). First and foremost, people seek jobs that optimize those skills, tasks, preferences, and interpersonal milieus that represent the best fit with their personalities. A good "fit" allows an individual to achieve a sense of well-being and adequate functioning in the work environment. Second, people seek work environments in which their personality type or compatible personalities predominate. That is, people want to be part of an interpersonal environment that enhances, and is compatible with, their own personalities. And there is no better way to achieve this harmony, according to Holland, than to find a work environment that is dominated by people with similar, or complementary, personality types. Holland is simply echoing the old adage that "birds of a feather flock together."

A core assumption in Holland's conceptual framework is that the stereotypes that people have about various occupations are fairly accurate. Although one may not know the specific day-to-day realities of a job, Holland cites studies that suggest that beginning in about high school, and independent of gender, age, or background, people develop opinions about broad occupational groups that remain fairly stable over time. He writes:

> Our everyday experience has generated a sometimes inaccurate but apparently useful knowledge of what people in various occupations are like. Thus we believe that carpenters are handy, lawyers aggressive, actors self-centered, salespeople persuasive, accountants precise, scientists unsociable, and the like. In earlier years, social scientists were skeptical of the amorphous folklore of vocational stereotypes (some still are), but recent work makes it clear that many have some validity (Holland, 1997, p. 9).

Holland concludes that: "If perceptions of occupations had no validity, interest inventories would have little or no validity, and the average person would have great difficulty in selecting suitable jobs" (p. 10).

Holland begins his theory of vocational personalities by asserting that parents tend to reproduce their own personality types in their offspring (p. 17). However, he is quick to point out that this is not a straightforward case of biological or psychological determinism. Instead, he assumes that parents provide "a large cluster of environmental opportunities," which goes beyond the parents' attitudes (p. 17). Rather, parents play a hand in their child's personality development because, by virtue of their own personality types, the parents are predisposed to engage in certain child-rearing practices that reward (or discourage) the development of some child personality traits over others.

Furthermore, parents tend to "surround themselves with particular equipment, possessions, materials, and tools"; to select friends and neighborhoods; and to engage in characteristic activities in and around the home that are expressions of their personalities (pp. 17–18). Thus, parents' attitudes are associated with a wide range of "environmental experiences" that shape and mold the child's personality (p. 17). And while children are being influenced by their physical and social environments, and by interacting with their parents, their personalities are also influenced by the mental and physical endowments inherited from their parents. But the development of personality is also textured by the fact that children also have a hand in creating their own environments. In the family, this reciprocal influence occurs to a limited degree through the demands that children make upon their parents and through parents' reactions to their children (p. 18).

Holland summarizes the process of vocational development well in the following paragraph.

> A child's special biology and experiences first lead to preferences for some kinds of activities and aversions to others. Later, these preferences become well-defined interests from which the person gains satisfaction as well as reward from others. Still later, the pursuit of these interests leads to the development of more specialized competencies as well as to the neglect of other potential competencies. At the same time, a person's differentiation of interests with age is accompanied by a crystallization of correlated values. These events—an increasing differentiation of preferred activities, interests, competencies, and values—create a characteristic disposition or personality type that is predisposed to exhibit characteristic behavior and to develop characteristic personality traits, attitudes, and behavior that in turn form repertoires or collections of skills and coping mechanisms (pp. 18).

The collection includes: (a) self-concepts; (b) perception of the environment; (c) values; (d) achievement and performance; (e) differential reaction to environmental rewards, stress, and so on; (f) preference for occupation and occupational role; (g) coping style; (h) personal trait; and (i) the repertoires of "skills" formed by a-h (p. 18). Thus, through a complex process of biological, environmental, and social influences and interactions, people develop activities, interests, competencies, and dispositions, which can be seen as a system of skills that signal an individual's personality repertoire.

Holland's personality types are "models of six common clusters of personality or behavioral repertoires that occur in our society" (pp. 20). The six types, and their parallel work environments, are realistic, investigative, artistic, social, enterprising, and conventional (pp. 21–28).[4] The distance between the personality types and their respective work environments is shown in Holland's hexagon (p. 36) in Figure 1.1, below.

The hexagon is said to provide a calculus for his theory that readily shows the amount of congruence or incongruence among types. More importantly, however, because each personality is said to be represented by a parallel work or occupational environment, the model is ultimately used to assess the degree to which a person's vocational personality fits his or her present or prospective job. For example, a person with an investigative

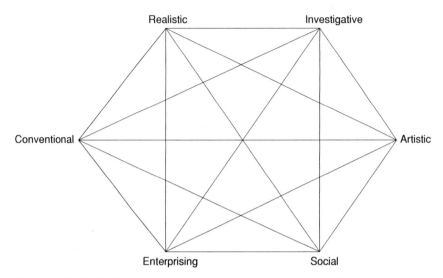

**Figure 1.1.** Holland's Hexagon of the Proximity and Interactions between Vocational Personality Types and Their Associated Work Environments

personality would experience a high degree of incongruence in an enterprising work environment requiring a high degree of collaboration and speculation. However, an investigative person would be expected to experience less incongruence in a realistic environment, even though a "perfect fit" between personality and environment would still not be achieved.

### Social Psychological Contributions to Occupational Choice

A pathbreaking article on occupational choice, explicating an occupational choice model from a decidedly social psychological perspective, was written nearly 50 years ago by Blau et al. (1956). As mentioned earlier, Blau and associates (1956) were keenly aware of the need to understand the psychological processes that lead a person to choose one career over another. But they were equally aware that—personal preferences and propensities aside—all career choices are made within a larger sociohistorical context. The broader contexts in which we are located limit and shape our options, and thereby give rise to opportunity structures. Thus, Blau is an ardent opponent of psychological reductionism which attempts to account for all group or social behavior in terms of psychological principles. Instead, in this article, we get an early view of career choice as arising out of the interaction between psychological characteristics and what Blau (1964) eventually described as emergent social and structural properties, which may not be reduced to individual psychological characteristics, such as those presented by Holland.

Blau and associates take a multidisciplinary perspective on occupational choice by presenting a conceptual scheme which takes the following influences into account: psychological (e.g., personal attributes), economic (e.g., market conditions), and sociological (e.g., social experiences and social structure). Like Super, Holland and others, Blau and associates view occupational choice as a developmental process. They write:

> There is no single time at which young people decide upon one out of all possible careers, but there are many crossroads at which their lives take decisive turns which narrow the range of future alternatives and thus influence the ultimate choice of an occupation. Throughout, social experiences—interactions with other people—are an essential part of the individual's development. The occupational preferences that finally crystallize do not, however, directly determine occupational entry (Blau et al., 1956, p. 532).

Blau and his colleagues differ sharply with Super and Holland by insisting that occupational "choice" must be viewed from the perspective of both the job candidate and the prospective employer, who must deal with factors far outside the purview of any particular job candidate when filling an existing position or when creating a new one. Thus, regardless of what a person may want in terms of a job or a career, factors beyond his or her control, or even knowledge, must be taken into consideration when trying to explain why people end up in different occupations (pp. 532–533). As such, the processes of choice (of job) and selection (of job candidates) must be placed in a larger conceptual framework if one has any hope of explaining why certain people end up in particular occupational positions. The central factor which drives this process—on both sides of the equation—is the social structure (p. 533).

The social structure is shown by Blau and associates to have a twofold effect on occupational choice (p. 533).[5] On the one hand, social structure influences the personality development of the job choosers; while on the other hand, it "defines the socioeconomic conditions in which selection takes place" (p. 533). They are quick to point out, however, that the two effects do not occur simultaneously. At any choice point in an individual's career, the interests and skills which influence a decision have been affected by the past social structure, whereas occupational opportunities and requirements for entry are determined by the present structure.

The social structure gives form to personalities by shaping and molding our "biological potentialities" through, among other things, socialization patterns, educational processes, family influences, and access to economic resources (p. 533). The initial molding of these biological potentialities results in diverse individual characteristics which may directly affect occupational choice. These individual characteristics or "sociopsychological attributes" include general level of knowledge, abilities and educational attainment, social position and relations, and orientation to occupational life. Orientation to occupational life refers to career aspirations, the importance of a job, and the identification of job incumbents as role models. The sociopsychological attributes are important in an occupational choice model because it is through them that the immediate determinants of occupational choice arise. These immediate determinants of job choice include job information, job qualifications, perceptions of the social characteristics of a job, and the reward structure associated with a job.[6]

The social structure, as already noted, not only influences personality development, it also defines the socioeconomic conditions in which selection of a job and selection of a job candidate take place. According to Blau and associates, the social structure (consisting of, for example, the social stratification system, cultural values and norms, demographic characteristics of the labor force, the type of economy, and the level of technological development) is shaped, modified, and molded by historical changes, such as trends in social mobility, shifts in industrial composition, historical development of social organizations, and changes in the level and structure of consumer demand. Over time, the historical changes result in particular types of socioeconomic organizations present at any given time. It is the particular type of socioeconomic organization which gives rise to the immediate determinants of occupational selection from the selecting agency's perspective.[7]

It should be stressed that near the final point of the choice and selection process, where the immediate determinants of occupational choice and selection reside, a sort of cross-fertilization of influences occurs. For example, while an individual's occupational choice is directly influenced by such things as the information he or she possesses about a job and his or her technical qualifications, the individual's choice may also be influenced by some of the immediate determinants that are usually associated with the selecting agent. Blau and associates suggest that this might occur when the individual's initial job preferences are adjusted upon being confronted by the actual demand for such a position in the labor market, or when the actual amount or type of rewards associated with a particular job become known to the individual (pp. 535–536). Similarly, the immediate determinants normally associated with the individual may also influence the decision of the selecting agent seeking to fill a position. This might occur when initial experience or educational requirements are raised or lowered after the supply and quality of potential candidates have been assessed.

## *Structural Contributions to Occupational Choice*

Granovetter (1974) examines occupational choice within a social networks framework that is informed by labor market theory and economic notions of rational behavior. While insisting on the "imperfection of information" (p. 26), Granovetter accepts economists' axiom that behavior is rational. He shows that job information is a valuable commodity (p. 99) that is most efficiently and rationally exchanged through particularistic, personal contacts (p. 98) that can effectively screen out unpromising and redundant information. The notion of using personal contacts to obtain costly information sometimes flies directly in the face of conventional sociological assumptions about formal rationality, which posits that modern industrial societies increasingly allocate jobs through open, universalistic, impersonal, and meritorious actions (Murray, Rankin, and Magill, 1981, p. 119).

Labor market theory generally holds that labor is a commodity that is subject to the same forces that affect any other commodity: namely, employers buy labor and employees sell it on an equilibrium-seeking open market. Price is analogous to a worker's total wage and benefits package (Granovetter (1974, p. 25). The supply and demand of labor constantly ebbs and flows as the market tries to establish its point of equilibrium: the point at which the price of labor ceases to fluctuate according to supply and demand, whereupon an equilibrium price is found which establishes the market conditions for labor (p. 25).

In this economic view, information as well as rationality are assumed to be "perfect," so that everyone has access to the same labor market information and all actions are made on the basis of economic gain. Granovetter (1974), however, found important contradictions to the economic view of labor supply in his empirical data. For instance, he found that people "least involved in [job] search behavior are over-represented in jobs of the highest income level" (p. 38). Economic theory would say, to the contrary, that as wage increases, competition, and thus job searching behavior, should also increase so that incumbents of the highest paying jobs would have actively sought and competed for such jobs.[8]

Granovetter (1974) summarizes his basic criticism of economic theory's shortcomings in accounting for occupational choice (or rather job searching behavior) as follows:

> If price does not match men with jobs, we are left with the question of what does. The argument of the present study is that the relevant factors are social; that job-finding behavior is more than a rational economic process—it is heavily embedded in other social processes that closely constrain and determine its course and results (p. 39).

These social processes are determined by personal contacts—placed within larger social networks—that may lead to newer and hopefully better jobs. The personal contacts are "*family-social contacts*" and "*work contacts*." Family-social contacts include relatives, friends, and the friends of one's family (p. 41). Work contacts, on the other hand, consist of professional acquaintances, people with whom one works, teachers, and current or former employers (p. 41). The two types of personal contacts are structurally distinct from each other and can be most readily differentiated by the strength of the interpersonal ties which comprise them. Briefly, the strength of an interpersonal tie between two contacts can be gauged by a combination of the following elements (Granovetter, 1973, p. 1361): (1) the amount of time devoted to the tie, (2) the suffusion of emotional intensity in the tie, (3) the amount of intimacy and mutual confiding in the tie, and (4) "the reciprocal services which characterize the tie." Family-social contacts are "strong ties," marked by a high degree of intimacy, sharing, and mutual friendships that are reminiscent of Cooley's definition of primary group affiliations—"intimate face-to-face association and cooperation" (Cooley, 1909, p. 23). Work contacts, in contrast, are "weak ties" which are not as strong, intimate or long lasting as the ties associated with family-social contacts, but which still pack more power than ties that are simply "nodding relationships."

From a network perspective, weak ties are of pivotal importance because in an information network, at least, they are the only relationships that can form effective links or bridges between people in different networks (Granovetter, 1973, p. 1364). Thus, the purposeful or serendipitous mobilization of a weak-tie contact can give an individual access to a whole new cluster of job tips, referrals, and contacts potentially valuable in finding and landing a new or different job. Weak tie contacts, moreover, tend to increase with occupational prestige. Strong-tie contacts, so the argument goes, are not as useful in acquiring new and useful information pertinent to a job search because the people with whom one is strongly attached are also likely to share the same contacts as oneself, thus reducing the probability of acquiring unique information that would be more forthcoming from a weak-tie contact (1973, pp. 1364–1365; 1974, pp. 41ff).

This argument is supported by empirical evidence (Granovetter, 1973). In a survey of several hundred men living in Newton, Massachusetts, Granovetter discovered that most of them found their current job in the following ways (adapted from data provided on p. 13):

- 56 percent through personal contacts
- 18 percent through formal means such as advertisements and employment agencies
- 19 percent through direct application
- 6 percent through some other means

Moreover, Granovetter found that those who used personal contacts had the better jobs, as gauged by objective standards. Among those who reported using a personal contact to find their current jobs, 69 percent said they used work contacts (of which 12 percent were teachers and 57 percent were work or professional contacts). Only 31 percent of these men relied upon family-social contacts to obtain their current jobs (pp. 41–42).

*Summary and Conclusions*

The central theme that arises from this section is that occupational choice is a multidimensional phenomenon that is part of a larger process. This multidimensionality is reflected in the context (i.e., "occupation") choice schema presented in chapter 3. In the case of the psychological theories, the larger process is centered on development of vocational maturity (via Super) and occupational personalities (via Holland). The larger process addressed in the social psychological schema put forth by Blau and associates is the emergent social and *structural* properties of society which have a powerful influence on the allocation of jobs and thus occupational "choice." In this view, occupational choices are constrained within a larger system of relations (e.g., social stratification system) that is outside the individual's control and of which he or she may be only partially aware. In this sense, Blau and associates' occupational choice model includes a key structural component not found in the work of Super and Holland. Finally, the larger process described in Granovetter's structural model of occupational choice is one's location in an information network which can relay nontrivial and nonredundant job tips and recommendations.

While the three broad approaches agree that occupational choice is a process, they are quite different in their basic presuppositions, especially when the psychological theories are contrasted to the two sociological theories. First, the theories of Super and Holland frame occupational choice within the context of psychological development. However, occupational choices are typically influenced by larger social organizations and processes. Both Super's and Holland's work may be considered reductionist because they attempt to account for occupational choice exclusively in terms of psychological phenomena. In my view, Holland's theory of vocational personality types is more reductionist than Super's vocational maturity indices, because the latter acknowledges much more interplay between one's personality and the social environment.

Relatedly, Holland and Super can be criticized for falling victim to the ontogenetic fallacy of adult development. Super tries to explain occupational choice by reference to age-related developmental tasks associated with five life stages. Holland, on the other hand, appears to discount the impact of social context and social structure in occupational choice, except insofar as individuals seek occupations and occupational environments that are compatible with their personalities. Holland makes a dubious claim when he says that work environments may be categorized into one of six typologies parallel to his personality types by discerning the dominant tasks and personality type operating within them. This assertion seems extremely tautological and sociologically insensitive to the complexities of work environments and social organizations. Furthermore, Holland believes that even if one has not taken a vocational preference inventory (as the Youth in Transition boys did not), one's vocational personality can still be assessed by knowing a person's vocational preferences, which were assessed. However, this does not seem like a reasonable claim, especially among high school and college students. As Super (1984, p. 226) says, in criticizing Holland, youths in high school or even college do not have enough knowledge of the world of work upon which to make valid assessments, and therefore informed occupational preferences.

Blau's social psychological model of occupational choice avoids psychological reductionism and the ontogenetic fallacy by focusing attention on occupational choice as a function of individual characteristics being molded, shaped and "marketed" within a larger social structure and historical milieu. Thus, one's personality, for example, is important but it is only one of many factors influencing occupational choice and recruitment. Moreover, Blau and associates show that the transition to work may be controlled more by external forces than by individual personal attributes. At the very least, structural and personal factors work together in occupational decision-making. Blau's scheme (he is adamant about not calling it a theory) is especially useful because it places occupational choice in a multidimensional framework that is sensitive to the contributions of both psychology and sociology.

Granovetter, finally, is far removed from both the psychological framework of Super and Holland, and from Blau's conceptualization. His chief interest is in understanding how social networks help or hinder occupational access and mobility, regardless of personality. He shows that the larger one's network of weak personal contacts (i.e., contacts with people outside one's own family-social network), the more likely a person is to obtain nonredundant, relevant information which may facilitate a job change. The psychological attributes of individuals in the network are of little or no importance, although their social characteristics (such as socioeconomic status) are important to understanding their position in network structures.

## Conclusion

This chapter has examined diverse life course and human development perspectives stemming from the fields of sociology and psychology. Elder and Rockwell (1977, p. 3) provide a good working definition of life course analysis which points to its interdisciplinary nature. According to them, life course analysis looks at the "pathways through the age-differentiated life span, to [the] social patterns in the timing, duration, spacing, and order of events and roles." Broadly conceived, life course research incorporates history, psychology, demography, and sociology (Hareven, 1981, pp. 144–145). Specifically, life course research deals with the complex issues surrounding constancy and change in human development, either explicitly or implicitly (Elder, Johnson, and Crosnoe, 2003).

In order to reflect the complexity of human development and life course issues (as well as their interdisciplinary nature), three broad areas of literature have been reviewed. First, key psychological and sociological perspectives on the life course, including hypotheses on the constancy and change of attitudes over the life course, were examined. While psychological theories (e.g., the work of Erikson and Levinson) have served as useful tools for understanding development, they fail to recognize socially produced change while favoring age-graded change tied to specific life cycle stages. Furthermore, advances in life course research have begun to seriously question the common psychological assumption that personality and attitudes crystallize in adolescence and early adulthood and remain relatively stable thereafter. Dannefer's (1984, p. 1084) sociogenic thesis criticizes the stage theories which are weakened by what he calls an ontogenetic fallacy. Glenn's (1980) aging-stability hypothesis and Elder, Johnson, and Crosnoe's (2003) paradigmatic principles of the life course, which incorporate developmental and environmental perspectives on constancy and change over the life course, were presented as an important theoretical perspectives from which to frame the substantive analyses presented in chapter 4.

The second area of literature concerned socialization to roles. This literature points out that social environments and organizations have an immediate and powerful impact on the individual by virtue of the socialization that takes place within them. This vast and vibrant literature is vital to framing both the empirical analyses in chapters 3 and 4 as well as informing the sociohistorical and cultural events that surrounded and help shape the life courses of the boys of the class on 1969.

Finally, the occupational choice literature was reviewed because it not only reflects the competing perspectives on occupational choice stemming from psychology and sociology, but because it also helps demonstrate the need to examine context choice from a multi-disciplinary perspective. The central focus of my study is not occupational choice per se, but the preliminary choice of the social contexts after high school—work, military, and college—which can heavily constrain occupational choice and attainments in the future. The findings from the occupational choice review (especially the insights of Blau et al. 1956) are used to frame the conceptual scheme used in the analysis of post-high school context choice presented in chapter 3.

Attention is now turned to the methods and procedures used to investigate the interrelations of contexts and self-concept change in the transition to adulthood.

# Notes

1. In sociology, the cutting edge of socialization research is socialization into roles. As used here, role is distinguished from status according to Linton's (1936) classic definition. Status is a socially assigned collection of rights and duties (Linton, p. 113). Thus, status is not necessarily an aspect of power as in bureaucratic authority, nor is it necessarily social esteem as in Weber's definition of status (1978, p. 305). In Linton's view, status is a structural variable which is both the sum total of an individual's rights and duties, as well as society's recognition of the various value and behavioral expectations to which the occupant of a particular status is expected to conform (p. 113). In societies based on entrenched statuses (i.e., the Indian caste system), role and status tend to converge. In societies with high social mobility (i.e., Western industrialized countries), the distinction between status and role is crucial. Role represents the "dynamic aspect of status" (Linton, 1936, p. 114). In short, a role is performed when the rights and duties of one's status are put into effect.

2. Super's application of vocational choice and adjustment to life stages is actually an extension of the pioneering work of Buehler (1933), who first delineated these five distinct stages of development. The stages are an outgrowth of her research on older men and women who were asked to reflect on what occupied and interested them at various times in their lives (Super, 1970, p. 435). Super does not discuss the childhood stage in any detail, focusing instead on the four latter stages beginning with adolescence.

3. The following is drawn from Burns' (1979, pp. 260–262) useful codification:

   1. *Growth stage (birth-14)*.

      The self-concept develops through identification with key figures in family and school; needs and fantasy are dominant early in this stage; interest and capacity become more important in this stage with increasing social participation and reality testing. Substages are:

      *Fantasy (birth-14)*. Needs are dominant; role-playing in fantasy is important.

      *Interest (11–12)*. Likes are the major determinants of aspirations and activities.

      *Capacity (13–14)*. Abilities are given more weight, and job requirements (including training) are considered.

   2. *Exploration stage (age 15–24)*.

      Self-examination, role tryouts and occupational exploration take place in school, leisure activities, and part-time work. Substages of the exploratory stage are:

      *Tentative (15–17)*. Needs, capacities, values and opportunities are all considered. Tentative choices are made and tried out in fantasy, discussion, courses, work, etc.

      *Transition (18–21)*. Reality considerations are given more weight as the youth enters the labor market or professional training and attempts to implement a self-concept.

      *Trial (22–24)*. A seemingly appropriate field having been located, a beginning job in it is found and tried out as a life work.

   3. *Establishment stage (25–44)*.

      Having found an appropriate field, effort is put forth to make a permanent place for oneself in it. There may be some trial early in this stage with consequent shifting, but establishment may begin without trial, especially in the professions. Substages in the establishment stage are:

*Trial (25–30).* The field of work assumed to be suitable may prove unsatisfactory, resulting in one or two changes before the life work is found or before it becomes clear that the life work will be a succession of unrelated jobs.

*Stabilization (31–44).* As the career pattern becomes clear, effort is put forth to stabilize, to make a secure place, in the world of work. For most persons these are creative years.

4. *Maintenance stage (45–64).*

Having made a place in the world of work, the concern now is to hold it. Little new ground is broken, but there is continuation along established lines.

5. *Decline stage (65 and on).*

As physical and mental powers decline, work activity changes and in due course ceases. New roles must be developed; first that of selective participant and then that of observer rather than participant. Substages of this stage are:

*Deceleration (65–70).* Sometimes at the time of official retirement, sometimes late in the maintenance stage, the pace of work slackens, duties are shifted, or the nature of the work is changed to suit declining capacities. Many people find part-time jobs to replace their full-time occupations.

*Retirement (71 on).* As with all the specified age limits, there are great variations from person to person. But complete cessation of occupation comes for all in due course, to some easily and pleasantly, to others with difficulty and disappointment, and to some only with death.

4. Holland's (1997, pp. 21–28) personality types are depicted below. A separate depiction of the six parallel work environments is unnecessary because they are simply coextensive reflections of the personality types.

1. *The Realistic Type:* Persons in this category enjoy manual work that involves manipulating objects and working with tools, machines, or animals. They see themselves as being mechanically and athletically inclined. Being a surveyor or a mechanic typify this type.

2. *The Investigative Type:* Investigative personalities enjoy work that involves the application of math and science. They see themselves as being scholarly and intellectual but not very sociable. Being a chemist or physicist are examples of this type.

3. *The Artistic Type:* Artistic personalities enjoy unstructured and creative activities involving art, music, or writing. They see themselves as being expressive, original, intuitive, nonconforming, and introspective. Being an artist or writer appeals to this type.

4. *The Social Type:* Social personalities tend to enjoy activities that involve training, educating or supervising others. They see themselves as helpful and understanding. Being a teacher or counselor are examples of the social type.

5. *The Enterprising Type:* Enterprising personalities like to manipulate and motivate others in the achievement of organizational and economic goals. They tend to see themselves as being aggressive, popular, and self-confident leaders with good speaking skills. Being a salesperson or an executive typify the enterprising type.

6. *The Conventional Type:* Conventional personalities prefer work that is ordered, explicit, systematic, and organized around prescribed organizational and economic goals. They see themselves as orderly and compliant persons with clerical and numerical ability. Being an accountant or a clerk typify the conventional type.

5. Although the term social structure has been used in many different ways in sociology (cf. Blau, 1975), Blau et al. (1956, p. 533) defined it in broad terms as the "more or less institutionalized pattern of activities, interactions, and ideas among various groups" (Blau et al., 1956, p. 533).

6. Several of these immediate determinants have very close parallels to factors cited by Super and Holland. For example, the accuracy of one's job information may be a function of one's vocational maturity (Super) and the stereotypic job characteristics we perceive (Holland). Our technical qualifications may be partially viewed as aspects of our repertoire of skills (Holland) and our location along the vocational development continuum (Super). Perceptions of the social aspects of a job seem to be very closely related to the attempt to fit the vocational personality to the prospective job environment (Holland). Finally, the reward structure of an occupation also would have direct appeal to our vocational personalities.

7. More specifically, the socioeconomic organizational climate (marked by occupational distributions and labor turnover rates, division of labor, and relevant organizational policies in the private and public sector) produces specific determinants of occupational selection such as formal demand for jobs, amount and type of occupational rewards, nonfunctional job demands which affect job selection but not performance, (e.g., veterans status, gender or race), and functional job requirements (e.g., technical qualifications).

8. Granovetter (1974) acknowledges that in contrast to blue-collar workers, it is often difficult to determine whether or not a professional, technical or managerial (PTM) worker is actually in the job market. This appears to be true because blue-collar workers are often forced to mount explicit job searches because of terminations, resignations, and lay-offs. Blue-collar workers also have much less free-time at their disposal during working hours than do PTM workers, thus militating against making discreet job inquiries, meeting contacts or simply "keeping one's eyes open" (p. 35). PTM workers, on the other hand, are in a much better position to learn about new job opportunities, whether solicited or not, because of the way their work time and their jobs are structured (pp. 35–37).

# 2

# The Youth in Transition Study
## *Data, Methods, and Measurement Issues*

### Data Source

In order to assess the impact of social contexts and objective statuses on the stability of self-esteem during the transition to adulthood, secondary data from Bachman's Youth in Transition study (YIT) is used (see Bachman, O'Malley, and Johnston, 1978 for a description of the study). YIT is a longitudinal study utilizing a representative sample of 10th-grade high school boys attending 87 public high schools in the continental United States in the fall of 1966 (p. 2). Field operations were conducted between 1966 and 1974 by the University of Michigan Survey Research Center.

### *Sampling Design*

Bachman et al. (1969, pp. 21–23) employed a multistage sampling design executed in three successive stages. In the first stage, 88 counties and metropolitan areas within the continental United States were randomly selected as the primary sampling units. Each stratum consisted of approximately two million people. One stratum was dropped, however, because no high school in the area was willing or able to participate in the project. During the second stage, one public high school was randomly selected from each of the remaining 87 strata defined in the first stage. Furthermore, the probability of a particular school's selection was proportional to the size of its 10th-grade male population (p. 3). This sampling methodology had the twin advantage of selecting approximately equal numbers of boys from each high school, regardless of its 10th-grade male population, while giving all boys an equal probability of appearing in the nationwide sample (Bachman et al., 1969, p. 22). While the sampling strategy was relatively expedient and cost-efficient, it also increased the accuracy and representativeness of the data by helping to ensure that boys from different sized high schools—and school environments—had a similar probability of being selected into the study. In the third stage, 25 to 30 boys from each selected high school were randomly selected and invited to participate in the study.

The YIT panel was eventually surveyed five times—three times during their high school years (1966 to 1969) and two times after most left high school (1970 and 1974). The first four waves consisted largely of group and individually administered interview data and questionnaires, while the fifth wave of data were collected entirely through mailed questionnaires (Bachman, O'Malley, and Johnston, 1978, p. 5). The overall attrition rate from Wave 1 (1966) to Wave 5 (1974) was 29%. Table 2.1, below, shows the number of panel members in each wave (unweighted and weighted), the seasons and years interviewed, and the cumulative attrition and refusal rates calculated on the basis of the unweighted target sample of 2,277 (adapted from Bachman, 1975, p. 2; Bachman, O'Malley, and Johnston, 1978, p. 5).

## Sample Weighting

A weighting scheme was developed by the Michigan survey staff in order to correct the sample by compensating for the selection of boys

**Table 2.1. Data Collection Points, Frequencies, and Attrition Rates: Youth in Transition Study Total Sample**

| Wave[a] | Grade | N (Unweighted) | N (Weighted) | Attrition (Cumulative) |
|---|---|---|---|---|
| Target sample (1966) | 10th | 2,277 | — | 0% |
| Wave 1 (Fall 1966) | 10th | 2,213 | 2,514 | 3% |
| Wave 2 (Spring 1968) | 11th | 1,886 | 2,158 | 17% |
| Wave 3 (Spring 1969) | 12th | 1,799 | 2,053 | 21% |
| Wave 4 (Spring 1970) | HS + 1 yr[b] | 1,620 | 1,853 | 29% |
| Wave 5 (Sp-Sum 1974) | HS + 5 yrs | 1,628 | 1,866 | 29% |

[a]Data were collected at school in Wave 1, at a neutral site in Waves 2 through 4 (i.e., church, library, community center), and through a mailed questionnaire in Wave 5. Respondents serving in overseas military bases, regardless of wave, were sent a mailed questionnaire.
[b]Not all subjects graduated from high school. This is short for Wave 3 plus 1 or 5 years.

from schools where the 10th-grade male population was smaller than the 25 to 30 boys normally sampled from a school (Bachman et al., 1969, pp. 126–127; 1978, p. 3). Three weights were used, with the average value across all respondents being 1.14 (1978, p. 3). In all, 299 boys were given double weights and one boy was given a triple weight; everyone else was weighted 1.0 (Bachman, 1970, p. 4).

Bachman correctly pointed out that sample weighting tends to increase sampling error. However, he argued that the trade-off between increasing sampling error versus decreasing systematic bias was justified because the resulting decrease in measurement accuracy due to weighting is quite small (1969, p. 127). Following Bachman's recommendations, I use weighted data, unless otherwise noted.

## Panel Biases

Two types of bias were explored by Bachman et al. (1978, p. 6): bias due to repeated measures and bias stemming from panel attrition are extremely important to consider when using longitudinal data. In order to test for repeated measures bias, a control group was drawn in 1966 in addition to the regular sample. The reason for drawing a control group was to test for subject sensitization to repeated measures.

The control group consisted of 115 subjects whose parents still lived in the same home as they had when their son was in 10th-grade (Bachman et al., 1978, pp. 257–261). The control group was contacted for the first time in April 1970 (during Wave 4 data collection) and interviewed at sessions held at public facilities similar to those used for interviewing the regular panel members (p. 257). The control group was asked the same questions as the regular panel (except for minor modifications to questions that referred to previous experiences in the study). In all, responses to 104 variables were systematically compared between the control group and the regular sample via t-tests and chi-square tests. Bachman et al. (1978, p. 6) concluded that no appreciable differences existed between the control and study group. Specifically, the two groups differed significantly on five variables at the .05 level of statistical significance; however, this number of differences could have occurred by chance alone (p. 258). The control group measured higher on flexibility, ambitious job attitudes, and the use of amphetamines, while measuring lower on resentment. People in the control group were also more likely to be high school dropouts (p. 258).

An examination of panel attrition, however, revealed some bias. Twenty-nine % of the original panel either refused to participate in the study or dropped out by the fifth and last data collection point in 1974, nearly eight years after their initial participation. Lin (1976, p. 241) notes that among those doing cross-sectional surveys, a 70% response rate is considered very good, and by inference, a complex longitudinal study such as the Youth in Transition project, with a subject retention rate of 71% across five data points spanning an eight year period, should be considered excellent.

Bachman et al. (1978, pp. 6–11, 258–264) found that there was a systematic, though in their view not serious, loss of subjects from low socioeconomic backgrounds, racial minority groups (especially blacks), among high school dropouts, and among persons in the lowest categories of general intelligence (pp. 6–11). Their conclusions were based on comparing differences between the study's droppers (n = 585) and stayers (n = 1,628) on selected univariate (mean difference), bivariate (product-moment), and multivariate (regression coefficients) statistics, using the unweighted sample (pp. 258–264). Still, relatively recent advances in correcting for sample selection bias now makes it possible to closely approximate the randomness of a previously random sample by including a "selection term" in one's substantive equation. The issue of selectivity bias is quite complex and is therefore dealt with extensively in a later section.

## Design Cluster Effects

The principal difficulty with multistage sampling designs, from the point of view of sampling theory, is that clustering reduces the number of independent observations which in turn increases the sampling error (Sudman, 1976, p. 76). Formulas for obtaining sampling errors in simple random samples are not appropriate for stratified cluster sampling designs such as YIT because they tend to underestimate actual sampling errors (Bachman, O'Malley, and Johnston, 1978, p. 253). Bachman et al. (1978), Sudman (1976, 1983), and Frankel (1983) provide some guidance for dealing with design effects; however, because of the subsample used in the dissertation, the use of such formulas is quite problematic. As such, I prefer to use standard error estimates that assume a simple random sample, while alerting the reader to the possibility that some standard error estimates may be underestimated. It should be noted, however, that even if the overall variance in a sample is compressed because of clustering, if one assumes that the bias is relatively uniform across all variables, then the clustering effect should be canceled out. Stated differently, if the variance in the sample statistics are proportionally constant through out the data, then one's parameter estimates (e.g., regression coefficients and so on) will probably not be seriously biased because the data dispersions will be uniform and will effectively cancel each other out. As such, I will use the standard alpha level of .05 to assess significance. The reader is simply reminded that there may be a slightly higher potential for committing Type I errors (i.e., rejecting a true null hypothesis) if a clustering effect is present in the sample.

## *"School Effects" on Minority Subjects*

Given the pre-busing era in which the boys from this study went to high school, the blacks who were sampled generally were located in only nine largely segregated high schools (Bachman, 1970, p. 5). As such, something that may be usefully referred to as a "school effect" exists for many of the nonwhite subjects, making their representativeness of the 10th-grade United States racial minority population questionable. Bachman (p. 5) observed this by writing:

> The limitations of the sample become more severe when we analyze subsets of the sample—particularly subsets affected by the school clustering. The most serious problem of this sort involves the black subset of our sample. While the number of black students in our sample (256—about 11% of the total) is fairly consistent with census data, the majority of these cases is located in only a handful of all-black schools. This is no doubt consistent with reality— most blacks do attend segregated schools. But given our sampling methods this means that our data on blacks are drawn from just a few clusters, and are thus subject to a great deal more sampling error than is true for our white respondents.

Given these important concerns about the representativeness of blacks, I have chosen to use only whites in this study, unless otherwise noted.

## Defining the Social Contexts

### *Initial Subject Selection Decisions*

Because of my substantive interest in the influence of three social contexts—work, military, and college—on self-esteem during the transition to adulthood, I initially chose to confine my analyses to only these three groups. However, while reviewing the issue of selectivity bias, I became convinced that the accuracy of my findings would be enhanced if I included all available persons in my sample. Therefore, all of my analyses were re-run to include a contextually heterogeneous "other" group consisting of everyone present in at least Wave 3 and Wave 5 who *were not* classified as being in the work, military or college social context groups.

Regardless of one's eventual social context classification, a person was included in my final sample if he met the following two conditions:

1. A person was required to be present in at least Waves 1, 3, and 5, although he may or may not have been surveyed in Waves 2 and 4. The Wave 3 data, recall, were collected in the spring of 1969 when most of the subjects were completing their senior year of high school, while the Wave 5 data were collected five years later, in 1974. The total number of whites who were present in Waves 1, 3, and 5 was 1,550 (71% of the 2,177 whites represented in the first Wave.)

2. Complete data were also required on all of the variables used in the substantive model of the effect of social context on self-esteem. The full substantive model reported in chapter 4 has 31 variables drawn from Waves 1, 3, and 5 and N of 1,164. (Although this represents a 25% drop in the total pool of 1,550 whites due to missing data, it is assumed that the data are missing randomly. This, and the possibly more serious issue of attrition bias, is discussed in detail in the section on sample selection bias later in the chapter. The methods by which the two self-esteem scales were created is described in detail later in this chapter.

## Criteria for Constructing the Social Context Groups

The rationale for examining how post-high school social context may influence later self-esteem was discussed in chapter 1. My goal now is to briefly review why the social contexts are hypothesized to be important to psychological development as manifested in self-esteem and to define exactly how the main social contexts of interest were constructed. To review, the three main social contexts of central interest to me are: (1) the full-time labor force, (2) the full-time Federal military (excluding the Reserves, National Guard or ROTC), and (3) full-time college. A fourth residual group consisting of subjects not otherwise classified was also constructed; this group is referred to simply as "other." The number of men falling in each context is listed in Table 2.2, below.

**Table 2.2.  Social Context Groups: Subjects Present in at Least
Wave 3 and Wave 5**

| Social contexts | N | Percent |
|---|---|---|
| Full-Time Worker (During At Least Waves 4 and 5) | 201 | 13 |
| Military Service > 18 Months Duty | 263 | 17 |
| College Graduate or > 16 Years Total Education | 479 | 31 |
| Other (Not Classified In Any of The Above) | 607 | 39 |
| Total | 1550 | 100 |

*Note:* Wave 4 data, on work experience, are only used to construct the full-time work group because complete job history information is unavailable. We assume that there is a very high probability that men who worked full-time in Wave 4 (1970) and Wave 5 (1974), never served in the military, and receive more than one year of post-secondary schooling were in the labor force during most of their post-high school years.

Two leading criteria are used to construct the three main social contexts: mutual exclusiveness and a major post-high school involvement in a context in terms of time or immersion. The two criteria are discussed briefly in turn. First, the main contexts are constructed so that categorization in one context excludes categorization in another. Only in the case of the military group is this criterion somewhat relaxed. Although a serviceman could have been a full-time worker in Wave 4 and Wave 5 (e.g., served in the military between the summer 1971 and winter of 1974, he is categoried in the military context if he served at least 18 months on active Federal military duty. Servicemen with 18 or months of active duty were also placed in the military group even if they received some post-secondary education before or after the service. However, if a man first went to college for four years and then entered the military, he is considered to have had a significantly mixed post-high social context experience, and is therefore placed in the "other" group. (Besides, it would be very unlikely for a YIT subject to have gone to college for four years *and* served 18 months in the military by Wave 5.) Men who served in the National Guard or military reserve were also grouped into the "other" category. I did this because the ongoing effect of involvement in the guard/reserve could be inextricably confounded with influences stemming from involvement in one of the other civilian contexts. (The guard/reserve requires a six year minimum enlistment. A guard/reserve recruit typically receives four to six months of initial full-time military training at a Federal military installation before returning to his community whereupon he attends monthly meetings and participates in an annual two week military training exercise.) None of the men categorized in the college or work group had any military experience, including guard/reserve duty.

The second criterion combines ideas about the quantity and the quality of the social context involvements. In terms of quantity, I suggest that the longer a person is involved in a particular social context, the more likely the context will have some kind of a significant psychological impact upon him. Thus, if the social contexts are conceptualized as social organizations encompassing complex systems of roles, norms, statuses, formal rules, and so on to which an individual is socialized, then the longer one is a member of a context, the more likely he will be influenced by it. Another aspect of quantity involves open- and close-endedness. On the one hand, being in the labor force is open-ended and encompasses most of a man's adult life; on the other, attending college or serving in the military are both typically close-ended involvements by design or custom. Thus, college is designed to be completed in four years, while military service is seen as a temporary duty for the majority of American men, who typically serve no more than four years.

Given existing records and the basic assumptions about quantity described above, men were categorized in the work group if they are assumed to have been mostly working for pay at a full-time job for at least four of the five years between 1969 (Wave 3) and 1974 (Wave 5). Similarly, the men in the college group are assumed to have been primarily college students for at least four out of the five years following high school. The time commitment for the military group was less stringent than it was for the other two groups. Instead of requiring four years of military service, which would distort the military sample by including only those who served what is typically a *maximum* first term of service, the military group consists of all men who served a minimum of 18 months on active military duty up to the fifth wave in 1974. Eighteen months of Federal military duty at the time of the study was considered to be a full term of service.[1] Also, because the vast majority of men who served in the military during the Vietnam War completed one term of service before leaving (U.S. Veterans Administration, 1977, pp. 15, 21), it is reasonable to regard 18 months of service as the lower bound of normal military duty.

Second, a qualitative aspect of experience is also an important criterion to use in composing the social contexts, although it is more subjective. By qualitative, I mean the nature of the socialization demands placed upon the individual by the context. Work, of course, demands certain things from an employee such as being dependable, conforming to the rules of the workplace, but most of all performing one's job with a minimum of competence. The workplace, however, is not usually thought of as primary socializing agent in terms of having an explicit organizational goal of so-cializing or re-socializing its members (see Gecas, 2003). Therefore, while the workplace may in fact be an important context in which adult develop-ment may continue through socialization, that is not its primary function. Quite the opposite is true about college and the military. As Wheeler points out (1966, p. 68), colleges and universities are "developmental socializa-tion systems" with specific organizational goals of training, educating, and socializing the individual. A person who has completed four years of col-lege, therefore, has been exposed to a full college "treatment" and has passed through a fundamentally qualitative socialization experience. The military can be thought of broadly as a "re-socialization system," according to Wheeler (1966, p. 68) which like other such organizations attempts to force its new members to abandon old identities, beliefs, and values "in the process of creating a new self-concept and world view" (Gecas, 1981, p. 168). As such, even if the men comprising my military context group were not involved in the context as long as the work and college people were, the very qualitative nature of their experience should be sufficient for classifying a man into the military context if he completed what is generally regarded as a minimum full term of service (18 months). There is, however, one further exception to the time-in-context requirement for the military veterans. If a man served in Vietnam but did not serve a full 18 months in the military, he was still classified in the military context. This exception is made because it is assumed that military duty in Vietnam is of such overriding importance that it marks a military experience that is sufficient powerful to justify including any Vietnam veteran in the mil-itary context regardless of length of service. And even though no data on the topic exist, it is reasonable to surmise that Vietnam Veterans with less than 18 months total service were released early because of illness, injury or disciplinary and performance problems. All of these would constitute noteworthy context-oriented events.

The exact criteria used to construct the three main social contexts are presented below.

*The Full-Time Work Group*

Defined by persons for whom all of the following applied:

1. They hold a full-time job in 1970 (Wave 4).
2. They were "mostly working on a job" in 1974 (Wave 5).

3. They were working at least 32 hours per week in 1974.
4. They never served in any branch of the military, including the National Guard or reserves.
5. They received no more than one year of total post-high school education.

They were allowed to have up to one year of schooling beyond high school in order to allow for experimentation with post-high school education or to allow for enrollment in a vocationally-oriented course of study (e.g., welding, heating and refrigeration). To the best of my knowledge, no workers embarked on a lengthy course of study while working full-time.

*Military Service Group*

Defined by persons for whom all of the following applied:

1. They reported active Federal military service in the Army, Navy, Air Force, Marines, or Coast Guard.
2. They reported serving 18 or more months on active Federal military duty.
3. A subject was placed in the "other" group if he reported serving in the military but did not provide his branch of service, which would make it impossible to know whether he served in the Federal military or in the guard/reserve.
4. If a person reported a start date for military service but no end date, he was assigned to the military group only *if* his service began on or before January 1973 *and* he reported being on Federal military service during Wave 5 survey. In this case, I could conclude that he must have served at least 18 months of active duty by the time of the Wave 5 survey.
5. Those who reported Vietnam service, regardless of the other rules, are included in the military context because of the assumed importance of this service as a military experience (see Table 2.3, below).

**Table 2.3. Military Context Group by Vietnam Service Subjects Present in at Leave Wave 3 and Wave 5**

|  | Vietnam service | No Vietnam service | Total |
|---|---|---|---|
| Military Context Group | 89 | 174 | 263 |
|  | 31.7% | 68.3% | 100% |

*Four-Year College Group*

Defined by persons for whom all of the following applied:

1. They reported that they either had earned a bachelor's degree or had received at least 16 years of total education by the time of the Wave 5 survey.
2. They never served in any branch of the military, including the National Guard or reserves.

## Overview of Personal Characteristics by Social Context Group

Attention is now turned to answering three brief questions: (1) What were the subjects mostly doing when surveyed in 1974? (2) What were their incomes? (3) What were their marital statuses? A more complete description of the subjects social characteristics is provided in the next chapter.

By 1974 (Wave 5), most of the YIT subjects reported their main activity as either "working on a job" (69%) or being a "student in college" (17%). All of the contexts except the full-time work group, however, had quite diverse 1974 activities profiles (see Table 2.4A). The work group's main activity, for example, was working (97%), with only a few reporting that they were unemployed (3%). (The work group's profile is single-dimensional largely because it was defined partly by persons reporting to be working full-time or temporarily laid off during the 1974 surveys.) People in the military and college social contexts appear to have the most diverse activities profiles. While approximately half of the military context was working in 1974, a little less than a third (29%) was still serving in

Table 2.4A. Percentage Subjects' Main Activities in Spring of 1974 By Social Context[a] (N = 1,538)

| Activity | Work | Military | College | Other | Total |
|---|---|---|---|---|---|
| | | | Social Context | | |
| Votech School | 0 | 2 | 0 | 1 | 1 |
| College | 0 | 10 | 35 | 11 | 17 |
| Working | 97 | 50 | 54 | 79 | 69 |
| Military | 0 | 29 | 2 | 2 | 6 |
| Unemployed | 3 | 8 | 3 | 4 | 4 |
| Other | 0 | 0 | 2 | 1 | 3 |
| Total | 100 | 100 | 100 | 100 | 100 |
| | (201) | (261) | (479) | (597) | (1538) |

[a] The original question was: "What one phrase describes what you are doing *mostly*? (1) Student in vocational/technical school; (2) Student in college; (3) Working on a job; (4) Serving in the military; (5) Unemployed; (6) Other (please specify)."

the military. Similar to the "other" group, only about 10% of the military group was in college. Among those classified in the college group, slightly more than half (54%) reported working as their main activity, while a little more than one-third reported that they were still mostly in college (35%). Interestingly, 6% of the college group reported doing something other than being in school, working, serving in the military, or the like. (Unfortunately the "other" responses were not coded by Bachman et al. and are thus unavailable for my analyses.) Finally, the vast majority of people categorized in the "other" context indicated that they were mostly working in 1974 (79%), although they reported doing a wide variety of things in the interim. Another 11% of the "other" group reported that they were mostly in college.

The "other" group had the most diverse 1973 income profile, with its members reporting among the lowest and highest incomes of all the subjects (standard deviation equals $3,993, see Table 2.4B, below). One % reported having no 1973 income, while another 1% reported incomes in excess of $25,000. The mean income for the "other" group was $7,151. The work group reported the largest mean income ($9,155), presumably because its members had more work experience and job seniority, and therefore higher salaries. The average income for the military group was $6,243, which probably reflects two things: the generally low military pay for those still in the service in 1973 and the generally low entry-level pay for

**Table 2.4B. Percentage Total 1973 Income by Social Context[a] N = 1,523)**

| 1973 income[b] | Social context | | | | |
|---|---|---|---|---|---|
|  | Work | Military | College | Other | Total |
| $0–$999 | 1 | 2 | 13 | 4 | 6 |
| $1,000–$3,999 | 5 | 17 | 44 | 14 | 23 |
| $4,000–$7,999 | 30 | 58 | 36 | 45 | 43 |
| $8,000–$10,999 | 44 | 16 | 5 | 26 | 20 |
| $11,000–$14,999 | 18 | 3 | 1 | 8 | 7 |
| $15,000–$24,999 | 3 | 3 | 0 | 2 | 2 |
| $25,000 Plus | 0 | 0 | 0 | 1 | 0 |
| Total | 100 | 100 | 100 | 100 | 100 |
|  | (199) | (258) | (470) | (596) | (1523) |
| Mean | $9,155 | $6,243 | $3,851 | $7,151 | $6,239 |
| Standard Dev. | $3,332 | $3,597 | $2,745 | $3,993 | $3,922 |

[a] The original question was: "How much did you earn from working (salary, wages, tips, commissions) in the entire year of 1973 (before taxes)? (01) $0, (02) $1–$999 ... (18) $35,000 and above."
[b] The original 18 income categories have been collapsed to aid presentation.

the dischargees who had only recently re-entered the civilian labor force. The college group's mean income of $3,851 was very low because most of them were either still in college (and thus not working full-time at high paying jobs) or were in low paying entry-level positions like some of their military counterparts. (It should be noted that the income figures for 1974 are not used because they have large amounts of missing data.)

The incomes reported for each group were significantly different from each other at the .05 level of statistical significance, using the Scheffe test for post hoc comparisons of group means. See the analysis of variance results summarized in Table 2.4B-1, below.

**Table 2.4B-1. Oneway Analysis of Variance of 1973 Total Income by Social Context (N = 1,523)**

| Source | Sum of squares | df | Mean square | F | $Eta^2$ |
|---|---|---|---|---|---|
| Between groups | 4,885,443,138 | 3 | 1,628,481,046 | 133.51* | .26 |
| Within groups | 18,528,091,503 | 1519 | 12,197,558 | | |
| Total | 23,413,534,641 | 1522 | | | |

*$p < .05$.

Table 2.4C, below, shows that by 1974 nearly half of the respondents were married (49%) and a small percentage were either divorced or separated (3%). Notice also that over two-thirds (68%) of the college group had never been married, while over two-thirds (71%) of the work group were currently married. The military group may have had a slightly higher probability of being divorced or separated (7%), but the difference could be due to chance.

Overall, the military group and the "other" group had similar marital status patterns. A one-way ANOVA using a dichotomous dependent

**Table 2.4C. Marital Status in 1974 by Social Context[a] (N = 1,526)**

| Marital status | Social context | | | | |
|---|---|---|---|---|---|
| | Work | Military | College | Other | Total |
| Single | 27 | 38 | 68 | 45 | 48 |
| Married | 71 | 55 | 32 | 52 | 49 |
| Divorced/Separated | 2 | 7 | 0 | 4 | 3 |
| Total | 100 | 100 | 100 | 100 | 100 |
| | (199) | (261) | (467) | (599) | (1526) |

[a]The original question was: "What is your marital status? (1) Single; (2) Married with one or more children; (3) Married with no children; (4) Divorced or separated with one or more children; (5) Divorced or separated with no children."

variable representing married or not married (i.e., $0 =$ single, divorced or separated; $1 =$ married) showed that the groups had different rates of marriage ($p < .05$). Furthermore, a Scheffe test of post hoc comparisons showed that all groups were different from each other, except military and "other." See Table 2.4C-1, below, for the ANOVA results.

**Table 2.4C-1. Oneway Analysis of Variance of 1974 Marital Rate by Social Context (N = 1,523)**

| Source | Sum of squares | df | Mean square | F | Eta$^2$ |
|---|---|---|---|---|---|
| Between groups | 24.87 | 3 | 8.29 | 35.41* | .07 |
| Within groups | 56.34 | 1522 | .2341 | | |
| Total | 381.21 | 1525 | | | |

*$p < .05$

## Measurement

### *Main Dependent Variable*

*Self-Esteem*

The ten self-esteem indicators draw on highly similar inventories by Rosenberg (1965) and Cobb, Brooks, Kasl, and Connelly (1966) (Table 2.5). Four negatively and six positively worded items denigrate and promote the self, respectively. All items are global or general self-esteem statements free from particular contexts (i.e., family, school, work) or referents (i.e., academic, social, physical).[2]

Later in the chapter, I will discuss in detail the dimensionality of global self-esteem and present my arguments for dividing it along its positive and negative evaluative dimensions.

### *Main Independent Variables*

Several independent variables are considered in the present research, with some having multiple indicators and while others are assumed to be "perfectly measured" with single indicators. My purpose here is to present the three theoretical constructs that are used as independent variables in the analyses that follow in the next chapter. The constructs are family of origin socioeconomic status (referred to henceforth as SES), intellectual ability,

**Table 2.5. Self-Esteem Indicators and Response Alternatives Used to Construct the Self-Worth and Self-Deprecation Scales**

| Variable Names | Source | Item wording[a] |
|---|---|---|
| I am of worth | Rosenberg | P1. I feel that I am a person of worth, at least on an equal plane with others. |
| I have good qualities | Rosenberg | P2. I feel that I have a number of good Qualities. |
| I do things well | Rosenberg | P3. I am able to do things as well as most other people. |
| I am not proud | Rosenberg | N4. I feel I do not have much to be proud of. |
| I have a positive self | Rosenberg | P5. I take a positive attitude toward myself. |
| I am no good | Rosenberg | N6. Sometimes I think I am no good at all. (In Rosenberg's original inventory it was: "At times I think I am no good at all"). |
| I am useful | Cobb et al. | P7. I am a useful guy to have around. |
| I can't do anything right | Cobb et al. | N8. I feel that I can't do anything right. |
| I do a job well | Cobb et al. | P9. When I do a job I do it well. |
| My life's not useful | Cobb et al. | N10. I feel that my life is not very useful. |

Response Alternatives: 1 = Almost always true, 2 = Often true, 3 = Sometimes true, 4 = Seldom true, and 5 = Never true

[a] P = Positively worded items used to develop the self-worth scale. N = Negatively worded items used to develop the self-deprecation scale.
[b] Bachman used this 5-point scale. Rosenberg used a 4-point scale which asked the respondents to strongly agree, agree, disagree, or strongly disagree with a given self-esteem statement.

and post-high school occupational prestige. The first two are exogenous background measures consisting of indicators gathered in 10th, 11th, and 12th grade; the latter is a Wave 5 measure of each subject's rank in the occupational prestige hierarchy as measured on by the Duncan occupational prestige scale (see Reiss, Duncan, Hatt, and North, 1961).

## Family of Origin Social Status

This construct turned out to be a problematic measure due to high amounts of missing data on Wave 2 father's occupational prestige and Wave 1 family income. However, this important variable can be salvaged. SES is a useful overall measure of the hierarchical position of a boy's family within the larger social stratification system of the United States. It is useful as a background indicator because it has been shown elsewhere that—among adolescents and adults—self-esteem tends to increase with social class position.

Under the general rubric of social stratification, the sociological literature has long been interested in prestige and status rankings, going back at least to Weber's seminal writings on status and status groups. Weber (1978) defined status as "an effective claim to social esteem in terms of positive or negative privileges," while pointing out that status groups are composed of a "plurality of persons who, within a larger group, successfully claim a) a special social esteem, and possibly also b) status monopolies" (1978, p. 305–306). A person's status, according to Weber, may arise from his or her style of life, formal education, social class origins, economic privileges, and occupational prestige (pp. 305–306). In the important monograph *Occupations and Social Status*, Reiss et al. (1961) developed their widely used socioeconomic status index (SEI)—a scale composed of information on income, education, and occupational prestige. The income and education variables were straightforward measures of annual income measured in dollars and education measured in years of formal schooling. However, the occupational prestige ratings, which were developed by Duncan, are much more complicated.

Duncan's occupational prestige ratings were developed by transforming North-Hatt occupational prestige scores—prestige scores obtained from interviews with American males in studies conducted at the National Opinion Research Center in the 1940s (Reiss et al., 1961, p. 139ff). The transformation scores were designed to reflect the socioeconomic characteristics of 446 occupations classified in the 1950 U.S. Census for the male civilian labor force. The scores consisted of weights for each occupation that were based upon the education, income, and age of persons occupying a particular occupation, as determined by the 1950 Census. Age was used as an indicator because of the distinct age distributions of many occupations (p. 133).

The Youth in Transition data set contains Duncan occupational prestige scores for both the subjects' fathers (measured during Wave 2) and for the subjects themselves (measures in Wave 5). The family of origin SES scale Bachman, O'Malley, and Johnston (1978) developed, however, was fairly non-traditional, and is not used by me. They used a family of origin SES scale which they call socioeconomic level. It consists of six equally weighted components: father's education level, mother's education level, father's occupational prestige (measured on the Duncan scale), number of rooms per person in the home, number of books in the home, and a checklist of material possessions owned by the family (p. 250).

The family of origin SES index I use is more parsimonious and traditional than the one used by Bachman. It is comprised of the same items used by Sewell and Hauser (1975, pp. 17–18) in a longitudinal study of high school boys making the transition to adulthood: father's and mother's educational attainment, father's occupational prestige (from the Duncan

SES Index), and the household head's income. Two major problems with the Youth in Transition data, however, hamper developing such an overall SES index: (1) there are large amounts of missing data on all relevant variables, but especially on family income and father's occupational prestige; and (2) even for those who have reported family income, it seems reasonable to expect that many of the high school boys—especially 10th-graders—would be unable to give accurate information about their parents' earnings. In fact, Sewell and Hauser anticipated the latter problem in their study of adolescents, and relied instead upon Wisconsin tax records for information on the family income and father's occupation. The suspicion that many of the boys in the Youth in Transition study did not possess key SES information is partly born out by examining the amount of missing data in the items which might comprise a Sewell and Hauser-type SES index (see Table 2.6, below).

Table 2.6.  **Missing Data on the Family of Origin Socioeconomic Status Index Items**

| SES items | Missing data Weighted sample Whites in Waves 3 And 5 (N = 1,550) | Missing data Full weighted sample Whites only (N = 2,177) |
|---|---|---|
| Father's education | 5% | 6% |
| | (85) | (143) |
| Mother's education | 4% | 5% |
| | (60) | (113) |
| Parents' mean education[a] | 2% | 2% |
| | (31) | (54) |
| Father's occupational prestige | 14% | 17% |
| Prestige Wave 2 (1967)[b] | (213) | (384) |
| Family income Wave 1 (1966) | 35% | 37% |
| | (544) | (825) |
| Family income Wave 3 (1969)[b] | 8% | 25% |
| | (126) | (545) |
| Duncan SES Scale[c] | 15% | 26% |
| | (234) | (560) |
| Parents' education plus father's occupational prestige | 14% | 26% |
| | (213) | (560) |
| Parent's education plus family income at Wave 3 (1969) | 10% | 26% |
| | (151) | (573) |

[a] Parents' mean education consists of the mother's education, the father's education, or both (where data were available).
[b] The figure in column two is inflated because it includes people who attrited after Wave 1 (i.e., were not asked the appropriate question).
[c] Duncan SES = Parent's education at W1 + Father's Occupational Prestige Duncan Scale) at Wave 2 + Family Income at Wave 3 (Listwise deletion of missing data).

Given the fact that family income and father's occupational prestige both contain high amounts of missing data, it is simply not prudent to use an SES construct which contains both measures, which would require the elimination of over 400 subjects via listwise deletion on the variable. Furthermore, the effect of using value substitution and estimation schemes to replace or compensate for missing data is not well understood and still quite controversial (for a full technical discussion see Anderson, Basilevsky, and Hum, 1983). As such, I will use only parents' education and Wave 3 family income as a single indicator of the social status ranking of each boy's family of origin.

## *Intellectual Ability*

The measure of intellectual ability is composed of z-scores from four intellectual ability or aptitude tests that were administered in the 10th grade. The tests are (Bachman et al., 1970, pp. 46–47, 63–69):

(a) *Ammons Quick Test of Intelligence.* The Quick Test is an individually administered intelligence test utilizing word recognition, but without requiring the respondent to be able to read or write. The interviewer read aloud 50 words in order of increasing difficulty to each respondent and had him indicate on a line drawing which picture best depicted the meaning of the word. The test lasted from six to 10 minutes.

(b) *Gates Test of Reading Comprehension.* The Gates Test is a group-administered reading comprehension survey consisting of 21 short passages placed in order of increasing difficulty. Each passage has two or three missing words in it which requires the respondent to select the appropriate insertions from a list of five possibilities. A total of 20 minutes was allowed to complete the test.

(c) *General Aptitude Test Battery-J*: Vocabulary (GATB-J). The GATB-J is part of a "well standardized multifactor test battery developed by the United States Employment Service for vocational counseling" (p. 68). The test is composed of 60 sets of four words each, with two of the words being either synonyms or antonyms. The respondent's task is to the select the appropriate pair of words that have either similar or opposite meaning (p. 47). The test lasted five minutes.

(d) *General Aptitude Test Battery-I*: Arithmatic Reasoning and Numerical Aptitude (GATB-I). The GATB-I consists of verbally expressed arithmetic problems which requires the subject to both understand the nature of the problem presented, and to solve it using arithmetic. Bachman et al.

write that the GATB-I "measures intelligence and numerical aptitude" (p. 68). The GATB-I takes five minutes to complete. Bachman reported that he used the vocabulary and arithmatic reasoning tests because they "show the highest factorial validity for general intelligence" (p. 68). The GATB scores were also found to have strong product-moment correlations (in the range .60 to .81) with other intelligence test scores (e.g., Army General Classification Test, California Test of Mental Maturity).

The standardized alpha coefficient for the intellectual ability construct (.82) demonstrates that it possesses acceptable internal consistency and is therefore believed to be a reliable measure (see Table 2.7, below, for additional details). The standardized alpha is reported because the test scores have been transformed into z-scores. Bachman et al. (1970, 1972, 1978), for reasons not made clear by them, chose to use only the first three tests listed above in their intellectual ability construct. This seems rather arbitrary because their construct ignores the role of arithmetic reasoning, which is an important indicator of intellectual ability or aptitude, while placing major emphasis on verbal and reading skills.

*Additional Exogenous Variables*

The substantive model presented in Chapter 4 includes a number of other control variables (e.g., attitudes toward war, educational aspirations, GPA). Since they were not found to be significant, I do not give them full treatment here.

*Selection Bias Term*

The substantive model presented in Chapter 4 includes a selection bias terms meant to control for attrition across the five waves of the YIT study.

**Table 2.7. Means, Standard Deviations, and Intercorrelations among the Variables Composing the Intellectual Ability Index (N = 1,550)**

| Test | Mean | Standard deviation | Intercorrelations | | | |
|------|------|--------------------|------|------|------|------|
|  |  |  | Q | G-J | G-R | G-I |
| Quick Test | 112.2 | 10.5 | 1.0 |  |  |  |
| GATB-J Vocabulary Test | 21.1 | 6.0 | .39 | 1.0 |  |  |
| GATES Reading Test | 38.2 | 4.1 | .61 | .56 | 1.0 |  |
| GATB-I | 10.9 | 3.0 | .56 | .47 | .63 | 1.0 |

Since this is an important and complex issue, I give it extensive treatment in the following section.

## Sample Selection Bias

Because 23% (n = 496) of the total weighted Wave 1 YIT sample of 2,177 whites was not present in the final wave, it is possible that the remaining Wave 5 sample (n = 1,550) is no longer random, which means that sample estimators may be biased (see Table 2.1, earlier in the chapter, for wave-by-wave frequencies). And biased sample estimators can cause serious research problems. The present task is to: (1) present a basic overview of this problem known as selectivity bias, (2) indicate where possible sources of bias in the YIT data lie, (3) propose a "selectivity model" by identifying relevant variables to be used in a selection bias equation, (4) review the formal specification of the selectivity bias problem, and (5) evaluate the proposed selection model.

### *Overview of the Selectivity Bias Problem*

Sample selection bias belongs to a family of methodological problems that I will simply call selectivity bias (see Mirowsky and Reynolds, 2000; Breen, 1996; and Berk, 1983 for useful overviews). Selectivity bias is widespread in the social sciences and can have far reaching effects on the generalizability of one's statistical inferences when one's parameter estimates are biased (see, for example, Mirowsky and Reynolds, 2000; Breen, 1996; Berk, 1983; Maddala, 1983). Even though effective sampling methods are a key sociological concern, strategies for dealing with selectivity bias after data have been collected have not found their way into widespread sociological practice. "This neglect," Berk (1983, p. 386) stated several years ago, "represents a major oversight with potentially dramatic consequences. More than external validity is threatened. Internal validity is equally vulnerable even if statements are made conditional upon the available data."

Moreover, because of nonresponse, mortality, and so on, no research methodologies are immune from possible selectivity bias: not experimental or cross-sectional or panel methods.

In survey research, for example, a previously random sample may become biased because of subject attrition, missing data, subgroup analyses, incidental selection of subjects, and so forth. Selectivity bias can also occur in quasi-experimental data when subjects are not randomly assigned to treatment or control groups. The effect is that unmeasured variables that might influence the dependent variable are not randomly distributed across treatment and control groups. Other reasons for selectivity bias in quasi-experimental data include subject self-selection into the experiment or experimental groups, non-random selection decisions made by researchers (or program administrators), *ex post facto* exigencies of the research such as attrition or incomplete data, and so on. Selectivity bias is not uncommon in evaluation research, where program administrators may assign or predispose clients to treatment or control groups, thus making randomization difficult or impossible. Randomization may be impossible when evaluators are asked to compare the relative effectiveness of different types of programs on some performance variables after the programs are underway.

Correction for selectivity bias can be viewed as an attempt to reestablish the representativeness of survey or experimental data, which, for a variety of reasons, is no longer representative of the population from which the initial random sample was drawn. This correction procedure is accomplished by introducing a "correction term" into one's substantive statistical model which compensates or controls for the probability that certain subjects were nonrandomly excluded from one's sample. The correction term is a variable itself and falls under the rubric of Heckman-type hazard models, popularly called a "hazard rate" or the inverse Mill's ratio (in reliability theory).

The problem of biased sampling distributions has been recognized for several decades (Berk, 1983, p. 390). According to Maddala (1983, p. 257), Roy (1951) was among the first persons to deal explicitly with the problem of "self-selection bias" in his research on career choice—i.e., why someone might choose hunting rather than fishing as a career. (Self-selection, as we shall see, is a specific type of selectivity bias in which choices made by the subjects directly or indirectly influence the composition of the sample.) Roy showed that career choice is a function of one's perceived productivity in a profession, which is a central issue of self-selection into a given career. The general problem of truncated distributions, however, has been grappled with since the turn of the century (Berk, 1983, p. 390). (Truncation generally involves missing observations on the exogenous and endogenous variables. Missing observations on only exogenous variables, however, does not affect selectivity bias.) It wasn't until the mid-1970s that econometricians first began to employ new methods for effectively dealing with the problem of self-selection bias (Maddala, p. 258).

The problem of selectivity bias can be conceptualized broadly as an omitted variable(s) problem. Viewed in this light, the issue centers around specifying a selection model that explains why subjects are systematically excluded (or included) from a sample, in the absence, usually, of any observed variables that might explain such selection. Once the model is specified, a hazard rate is obtained for each person in the sample. The hazard rate is then included as an exogenous variable in the substantive equation as a "selection term," thereby correcting for selection bias. (Note: "selection term" is used synonymously with "correction term.") The selection term can be thought of as a proxy variable whose effect incorporates that of all of the unobserved variables which might account for the probability of each subject being excluded (or included) from the sample. The absence of such an observed "selection" variable makes the estimation of hazard rates necessary and gives rise to the notion that selectivity bias is an omitted variable problem.

The problem of selection bias might be clearer by reviewing the four models in Figure 2.1, below. The first three models do not produce biased estimates, but the last one does. I will start with a model composed of these three variables:

$Y$ = a performance variable (i.e., self-esteem)
$G$ = a dichotomous variable indicating presence in the Wave 5 sample ($G = 1$) or absence from the Wave 5 sample ($G = 0$)
$Z$ = Set of explanatory variables.

Now, suppose that we ask this important question: What happens if a Z is omitted (or not observed)? The models are assumed to represent the true relationship among the variables, whether omitted or not.

Problems of selectivity bias can be classified into three areas: (1) truncated regression models, (2) censored regression models, and (3) dummy endogenous variables (Maddala, 1983). Truncated and censored regression models comprise the two main kinds of biased samples, involving what econometricians call samples with limited dependent variables (Kennedy, 1998, p. 193). Each bias problem is reviewed in turn.

In truncated regression models, the observations of X (the exogenous or independent variables) and Y (the endogenous or dependent variable) are obtained only if the value of Y is above or below some threshold (Maddala, p. 5). Stated another way, values of the exogenous variables are known

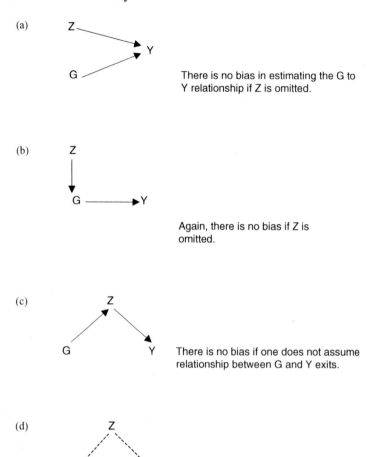

(a)

There is no bias in estimating the G to Y relationship if Z is omitted.

(b)

Again, there is no bias if Z is omitted.

(c)

There is no bias if one does not assume relationship between G and Y exits.

(d)

Bias is present in G to Y relationship if a Z is omitted (dotted line).

**Figure 2.1.** Potential problems when a variable is omitted

only when the dependent variable is observed (Kennedy, 1998, p. 194). Therefore, a case is selected contingent upon the person's score or value on the *dependent variable*, or sometimes on the way the observations have been aggregated because of the value of the dependent variable. Samples that are missing observations above or below some threshold are called distributions with upper or lower truncation (Maddala, 1983, pp. 366–367). There are also samples with upper and lower truncation, which Maddala calls doubly truncated distributions (p. 367). For example, the researcher may select observations on the basis of low scores on an intelligence test,

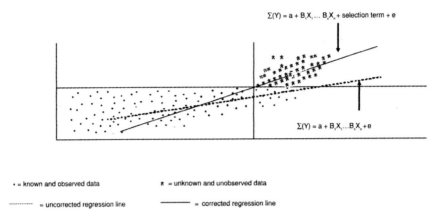

**Figure 2.2.** Data involving truncation and censoring: the limited dependent variable problem

because of high occupational prestige or because of low income. If one attempted to predict the value of a dependent variable from these samples through ordinary least squares regression, the dependent variable would be truncated (would not reflect the full range of values in the population) and one's estimates would therefore be biased (Kennedy, 1998, p. 189ff; Maddala, 1983, p. 3; Berk, 1983, p. 391;). In Figure 2.2, below, a sample is truncated if the X and Y values associated with the starred data have not been observed. In the illustration, Y values above $t_1$ and X values above $t_2$ are unknown and therefore missing from the sample. If the starred values were known, or if a selection term was included in the expectation equation, then the true or corrected (solid) regression line would have a steeper slope than the uncorrected (dashed) line.

Bias stemming from truncation would be most prevalent in surveys of special populations or in program evaluations where participants have extraordinary characteristics or needs (e.g., military rejects, people living below the poverty line, people with high occupational status, and so on). Maddala (1983) illustrates the truncation problem with the example of the difficulty encountered if one attempts to estimate an 'intelligence' equation from a sample of persons rejected from military service because of low military IQ test scores (based on the Armed Forces Qualifying Test [AFQT]). In this case, one might try to estimate AFQT scores as a function of education, age, SES, and so forth (1983, p. 3). The dependent variable, however, is truncated—part of the sampling distribution involving higher AFQT scores is missing—which can result in biased least-squares parameter estimates. Truncation also means that values for the exogenous variables in the equation may also be missing. If we select military rejects, for example, we end up with a truncated dependent variable (say AFQT), but we may also exclude recruits with higher educations and higher family SES's. The critical issue to remember is that a sample suffering from upper or lower truncation on the dependent variable will probably produce biased estimators.

Censored regression models involve situations in which all observations on X (the independent variables) are available, but some data on Y are missing. Put another way, censoring involves a selection process that eliminates observations solely for the endogenous variable (Berk, 1983, p. 391). In this case, actual values on Y are observed for some values of X, but for other values of X one only knows whether or not the absolute value of Y is above or below a certain threshold (e.g., zero or not zero). This type of sample has a distribution with right- and left-hand censoring. Maddala illustrates this by examining the case of married women in the labor force (p. 4), the problem that initially aroused the interest of econometricians in self-selection bias. Here, observations on Y (wage rate) are available (i.e., nonzero) only for women in the labor force, while there are no observations on Y (i.e., Y is zero) for women who are not in the labor force. Still, there are two categories of women—women in the labor force (who have jobs and therefore wage rate data) and women who are not in the labor force (and of course do not have jobs or wage rate data). Based upon economic theory, one assumes that a woman who does not participate in the labor force chooses to do so because her valuation of her time in the household (i.e., her reservation wage) exceeds her valuation of her time spent at a job (i.e., her wage rate). In Figure 2.1, above, censoring occurs when the Y values associated with the starred data points above $t_1$ are unknown or unobserved, but their corresponding X values are known and observed. As we shall see, I have identified censoring as a possible problem in the Wave 5 sample.

The problem of dummy endogenous variable bias was touched on indirectly in the example of a distribution with right- or left-hand censoring. In this situation a previously drawn sample is categorized and analyzed on the basis of some attribute, membership, or affiliation, defined by a dummy variable. This procedure can create a self-selectivity bias if the presumably dummy exogenous variable is itself *endogenous*—that is, if it is explained or accounted for by other variables in the model—then selectivity bias may occur. The dummy endogenous variable problem described here is really a model misspecification problem. Maddala (1983) provides the following example of dummy endogenous variable bias (p. 6).

> Suppose we are given data on wages and personal characteristics of workers and we are told whether or not they are unionized. A naive way of estimating the effects of unionization on wages is to estimate a regression of wages on the personal characteristics of the workers (age, race, sex, education, experience, etc.) and a dummy variable that is defined as
>
> $D = 1$ for unionized workers
> $D = 0$ otherwise
>
> The coefficient of D then measures the effect of unions on wages. Here the dummy variable D is exogenous. However, this is not a satisfactory method for analyzing the problem because the dummy variable D is not exogenous but endogenous. The decision to join or not to join the union is determined by the expected gain.

This, then, would represent a self-selectivity problem because the data are generated by self-selection of individuals into a labor union.

Not uncommonly, multiple sources of selectivity bias—on substantive and methodological grounds—operate in a sample. As we shall see in the next section, I have identified four sources of possible selectivity bias operating in the YIT sample.

*Possible Sources of Sample Selection Bias in the YIT Data Set*

There are four sources of possible bias operating in the Youth in Transition data set—bias due to initial participation refusals, attrition from wave-to-wave, missing data, and subject selection (e.g., the decision to exclude nonwhites because of the "school effect" described earlier). I will focus on the two main sources of potential selectivity bias operating in the data set: (1) bias due to subject attrition from Wave 1 to Wave 5 (i.e., attrition bias), and (2) bias due to loss of subjects through missing data (i.e., missing data bias). (Subject sensitization, a different type of bias in longitudinal studies, was discussed earlier). Figure 2.3, below, depicts the potential net effect of the four possible sources of selectivity bias in the original YIT sample.

The solid arrows ($\longrightarrow$) in Figure 2.3 point to the subjects remaining from the original sample after each round of eliminations. The dashed arrows ($----\blacktriangleright$) designate the actual number of subjects eliminated after each round.

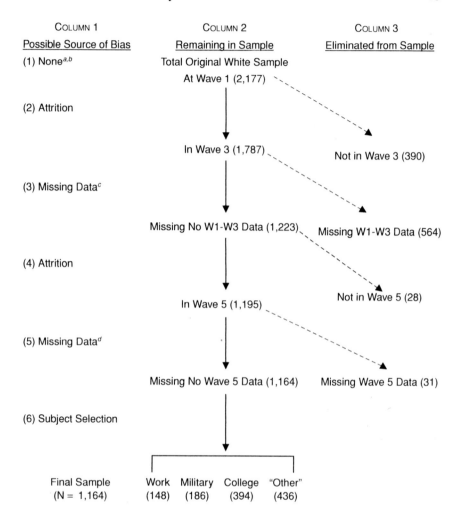

COLUMN 1 | COLUMN 2 | COLUMN 3

Possible Source of Bias | Remaining in Sample | Eliminated from Sample

(1) None[a,b]

Total Original White Sample At Wave 1 (2,177)

(2) Attrition

In Wave 3 (1,787) — Not in Wave 3 (390)

(3) Missing Data[c]

Missing No W1-W3 Data (1,223) — Missing W1-W3 Data (564)

(4) Attrition

In Wave 5 (1,195) — Not in Wave 5 (28)

(5) Missing Data[d]

Missing No Wave 5 Data (1,164) — Missing Wave 5 Data (31)

(6) Subject Selection

Final Sample (N = 1,164) | Work (148) | Military (186) | College (394) | "Other" (436)

[a] ⟶ cases remaining from the original sample
---▶ the actual cases excluded from the sample for the reasons given in column 1

[b] There are really two sources of potential bias not present in this line: (1) bias from boys who initially refused to participate in the study, and (2) potential bias stemming from excluding racial minorities. The proportion of eliminations due to initial refusals is so small (3 percent) as to be inconsequential. The possible selection bias due to the exclusion of nonwhites from the sample is not dealt with here because the presence of a "school effect" (discussed on page 75) is assumed to be more serious than confining the sample to whites only.

[c] Elimination on this line are based on missing data on any of the 28 variables used in the analyses of selectivity bias and context choice (see chapter 3).

[d] Eliminations on this line are based on missing data on any of the six variables composing Time 5 self-confidence.

**Figure 2.3.** Tree diagram of sources of selectivity bias for white sample (weighted data)

Column 1 of Figure 2.3 shows the three types of potential selection bias operating in the sample: attrition, missing data, and subject selection. I will use the terms selection (or selectivity) bias when referring to bias generally. When discussing a particular type of bias, I will refer to it by name. Bias due to refusing to participate in the study after having been selected in the initial sample is not considered to be a major problem in the YIT data set (see Table 2.1). Attrition bias may occur when subjects fail to appear in subsequent waves, for a variety of reasons discussed earlier. Missing data bias may result from the elimination of cases with missing data on the variables used in the substantive equations. Subject selection bias could enter into the final white sample if the "other" group (composed of the all the people who could not be categorized into one of the main contexts) was excluded from the final sample. More will said about this shortly. (The exclusion of nonwhites from the final sample is also a possible source of bias. I did not present this type of bias in the Figure 2.3 because I consider the school effect to be more of a problem than excluding blacks.)

Column 2 shows the number of subjects remaining from the original sample *after* each round of eliminations has been made. Column 3 shows the number of subjects eliminated from the original sample for the specified reasons. For example, line 1 of column 2 shows the total weighted white sample in 1966 ($N = 2,177$). Line 2 of column 3 shows that as of Wave 3, 390 people had left the original Wave 1 through attrition. The remainder of the figure follows the same general format.

The difference between lines 1 and 2 in Figure 2.3 is the initial source of potential selection bias; in this case, bias due to attrition from the Wave 3 sample. Line 2 shows that when attrition is taken into account, the original sample of 2,177 (line 1) was reduced to 1,787 subjects in Wave 3, because 390 subjects left the study (i.e., 18% of the Wave 1 sample.) Line 3 shows that of the 1,787 persons remaining in the Wave 3 sample, 564 of them had missing data on the relevant Wave 1 and Wave 3 variables used in subsequent analyses (particularly context choice analyses reported in chapter 3). This left 1,223 persons remaining in the usable Wave 3 sample (after listwise deletion of cases). Line 4 illustrates the further effect of attrition. Of the 1,223 people in the Wave 3 sample with no missing data so far, 28 dropped out of the Wave 5 study, which left 1,195 persons. After eliminating the 31 persons in the Wave 5 sample with missing data on the relevant Wave 5 variables (i.e., the self-esteem indicators), line 5 shows that 1,164 remained in the sample. Finally, line 6 shows the number of subjects remaining in each social context in the final sample after all of the various eliminations were made.

Subject attrition is probably the most serious source of bias in longitudinal studies. Attrition can directly alter the randomness of a previously random sample which in turn can significantly interfere with the integrity of the data set and ultimately lead to biased parameter estimates. Attrition bias in the YIT data would probably appear if the randomness of the Wave 1 data was significantly altered because certain subjects were systematically excluded from subsequent waves. Put differently, attrition bias increases as the likelihood of systematically losing or excluding subjects over the course of a study also increases. After reviewing the options, I decided that attrition bias should be the focus of my sample correction analyses.

As I mentioned before, Bachman found evidence that there was some, though not serious, systematic loss of subjects between Wave 1 and Wave 5 (Bachman, O'Malley and Johnston, 1978, pp. 6–11). He found that persons in the lowest categories of general intelligence and socioeconomic status, as well as racial minorities and high school dropouts, were more likely to leave the study than were others. This indicates that some selectivity bias in the form of attrition is operating in the data. Bachman based his conclusions on comparisons of stayers and dropouts on three levels (1978, p. 258): "univariate (do the means and distributions differ between the groups?), bivariate (do product-moment correlations differ between the groups?) and multivariate (do regression coefficients differ between the groups?)."

Referring back to the earlier discussion of censored versus truncated regression models, if I were to focus on attrition from Wave 3 to Wave 5, the problem would be essentially one of truncation. Truncation would occur because many values for X (as well as Y) would be missing for everyone who attrited up to Wave 3. For example, I would not have Wave 2 or Wave 3 explanatory variables for those who attrited after Wave 1. And Wave 3 dependent variables would also be missing for all persons who attrited. Focusing on attrition from Wave 1 to Wave 5, however, involves censoring. There would be complete data (for all practical purposes) for everyone in Wave 1, including a rich assortment of social, psychological, and demographic variables that might explain attrition. Therefore, we would have missing Y values (attrition at Wave 5) but would have complete X values stemming from the Wave 1 data.

The problem of measuring and then correcting for attrition bias can be dealt with effectively through a variety of statistical techniques including probit models, logistic models, and the more traditional linear probability models. I use probit modeling because it provides an effective and robust method for estimating the "hazard rates" for subjects who did not "survive"—for a wide range of known and unknown reasons—to the end of a study.

The second major potential source of bias may stem from missing data. Missing data bias is not usually thought of as a selectivity bias problem, but if we keep in mind that selectivity bias occurs in any situation in which there is a differential and systematic probability of excluding (or, conversely, including) cases in one's analyses, then selectivity bias may indeed operate when there are relatively large amounts of missing data, as in the Youth in Transition data set (e.g., 18% of the Wave 3 and 5% of the Wave 5 sample contain missing data).[3] If missing data is both sparse and random, however, one need not worry about its effect on the research results. However, if the proportion of missing data is large (above about 5–10% according to Anderson, Basilevsky, and Hum, 1983, pp. 452,467) or if it is nonrandom, then its potential biasing effect may be high. Because performing missing value estimates and the issue of substitution remains controversial, I have decided not to impute data, except in the initial selectivity bias models discussed below. Instead, a model-by-model listwise deletion strategy is employed so that each run includes all the available cases with complete data. The total N, therefore, may differ somewhat from chapter to chapter and procedure to procedure. But this is not an unusual practice, and is probably preferable to wading into the murky waters of either imputation or pairwise deletion.

Two major potential sources of sample selection bias have been identified, and solutions for dealing with them have been offered. Still, why not deal with all of the selection bias problems simultaneously and be done with them? This seems like a reasonable question, but is untenable for at least two reasons. First, an all-purpose hazard rate accounting for subjects lost through attrition and missing data would be uninterpretable. We would have no real idea of what the hazard rate actually referred to and this would severely limit its usefulness. Second, estimating two separate hazard rates and then using them in the substantive analyses is also not a viable solution. Some hazard rates will be partially based on earlier hazard rates, which opens the imminent possibility of biased hazard rate estimates. In short, basing later hazard rate estimates on earlier ones makes the error term highly biased because one estimate is being based on another estimate. The only tenable solution that seems open is to base the hazard rate estimates on attrition bias and deal with missing data through a modified listwise deletion procedure, as discussed above.

To summarize, there are two primary sources of bias in the YIT data set: (1) attrition prior to Wave 5, and (2) missing data in relevant Wave 1, 3 or 5 variables. One cannot at this time produce unbiased hazard rate estimates for both sources of bias at the same time. Consequently, attrition at or up to Wave 5 was identified as the main selectivity bias problem because it has the most direct impact on the randomness of the total sample. Second, the problem of missing data bias was solved in two ways. First, in order to have the selection model (from the Wave 1 data) be representative of the whole Wave 1 sample ($N = 2,514$), mean substitution was used in order to fill in missing values (this is discussed further below). Second, I decided not to substitute missing values in the substantive equations. Instead, each substantive equation includes all cases with complete data on the variables in that particular model. I dub this a modified listwise deletion strategy.

## Identifying Variables to Use in the Selection Equation

This part of the chapter discusses the choice of variables to be used in the selection equation—i.e., the model used to estimate the hazard rates. The hazard rate, recall, attempts to estimate the probability of remaining in or dropping out of Wave 5. Because attrition and retention are reflections of each other, the selection variables for the model are mostly framed in terms of what would influence someone to drop out of the sample.

When specifying the selection model, one must remember that the model should be constructed on the basis of theory—or at the very least on *a priori* grounds. If the exogenous variables included in the selection model are chosen from an *ex post facto* empirical examination of data from the same data set as the one suspected of being biased, as Bachman et al. did (1978, p. 257), then the empirical justification for choosing any exogenous variable could easily be subject to the very selectivity bias one is attempting to avoid.

The variables chosen for use in the selection equation are admittedly limited. On purely methodological grounds, they were selected with two considerations in mind. First, the variables are exogenous—that is, they are determined by causes outside my larger analytical scheme, and, epistemological arguments aside, the variables are not themselves explained by some third variable (under special situations, however, this assumption may be safely relaxed).[4] Second, the variables were measured in the first wave, so they apply to all subjects. (I did not use high school dropout status as an exogenous variable in the selection equation because it would require using the samples from Wave 2 and Wave 3, which may already be biased from attrition. Ideally, the variables used in the selection equation should be exogenous variables because endogenous variables can result in a biased hazard rate estimate due to model misspecification [this is the same logic that applies to the proper specification of causal models in general].) Consistent with general usage, however, this assumption is often relaxed. The main thing to avoid is an explanatory variable that is "simultaneous determined." That is, if the substantive dependent variable highly influences some attitude, behavior, or characteristic, and if that attitude, et cetera, is used in the selection equation, then the resulting hazard rate estimate will likely be biased. I have attempted to use some variables that, while not strictly exogenous, are not simultaneously determined. Admittedly, the list of selection equation variables is not exhaustive. However, if the model is adequately specified the resulting hazard rate will approximate the omitted variable for which it is a proxy, even if every conceivable variable is not included. More will be said about that later. For now, attention is focused on the actual variables used in the selection equation.

Little has been written about why people drop out of or remain in panel studies or what characteristics they possess. More is known about the difficulty of starting longitudinal studies, locating panel members in successive waves, securing their continued involvement, and so on (see, for example, Anderson, Basilevsky, and Hum, 1983, p. 417ff; Booth and Johnson, 1985). My own experience in survey research suggests that disinterest or hostility toward the survey's objectives or lines of questioning are two likely reasons subjects refuse to initiate or to continue participation. Lack of motivation is also a well-known reason for declining to participate or to remain in a study. Other reasons for dropping out might include subject geographic mobility (which can drive survey costs up regardless of whether a forwarding address was left), morbidity or mortality, and inaccurate or incomplete records which can make it difficult or impossible to locate subjects after they move.

We know that attrition has many sources and that it is difficult to anticipate and prevent; attrition is also very difficult to account for after the fact. Nevertheless, several background and demographic variables which may account for attrition between Wave 1 and Wave 5 were identified and included in a selection equation using probit modeling through the

LIMDEP 7.0 (Limited Dependent variable) program (Greene, 2002). After presenting the basic variable categories that were used, each selection variable is presented along with a brief description of it and my reason for including it in the model. These variables have been selected on theoretical, not empirical, grounds; that is, they were not selected on the basis of their *post hoc* empirical effect on dropping out of the study. While this may seem overly restrictive, recall that if one is using an empirical justification for including or excluding variables because of some observed differences between stayers and leavers, one may be basing those empirical results on a biased sample before the proper correction term has been applied.

There is one dichotomous criterion variable referring simply to whether or not a subject was present ($Y = 1$) or absent ($Y = 0$) in Wave 5. The 2,070 people represented in the dependent variable include only those whites who did not have any missing values on the eight exogenous variables used in the selection equation. The exogenous variables are grouped into four general categories: (1) four demographic variables, (2) one objective ability measure, and (3) one socioeconomic indicator, and (4) two school experiences variables.

### Selection Bias Dependent Variable

Wave 5 present (W5PRES): the endogenous variable signifying whether a subject was present in the Wave 5 sample ($Y = 1$) or was absent ($Y = 0$).

### Selection Bias Explanatory Variables

*Demographic Characteristics. Urbanicity*: a dichotomous variable composed of persons raised in rural areas ($Y = 1$)—e.g., farm, country but not farm, or small town, versus those raised in urban areas ($Y = 0$)—e.g., small city or large city. Urbanicity is included because rural youth might be more likely to leave the place in which they were raised in order to find employment, obtain job training or attend college, or simply move to a city, especially after leaving high school. This mobility may make them more difficult to locate and less willing to cooperate in further waves. (Moving did not exclude persons from the study.)

*Parents' Marital Status* (PARENTDIV): a dichotomous variable indicating whether the respondent's parents were divorced ($Y = 1$) at the time of the Wave 1 interviews or were married ($Y = 0$). This is assumed to be important because it approximates a "broken home" indicator. We might assume, for example, that children from broken homes will be more mobile and more difficult to locate upon moving than children in other families. Children of divorced parents may also be less willing to remain in a study that inquires about family relationships. Parental divorce in later waves, though interesting, is not included because it would require using data from subsequent waves, which are already marked by attrition.

*Number of living parents* (NUMBERPAR): a variable indicating whether any or all of the respondent's parents were alive at the time of the first interview. It may be important since children with a deceased parent may be more difficult to locate upon moving, burdened with financial responsibilities or other family hardships, and less willing or able to maintain interest in a longitudinal study.

*Working at Time 1*: a dichotomous variable indicating whether a boy reported that he was working at the time of the first interview ($Y = 1$) or not working ($Y = 0$). This variable is included because boys with early work experiences may have, or may develop, less interest in school. They may also be more likely to feel increased role strain resulting in conflicting school and work demands which may given them less motivation and less time or energy to devote to the study (see Mortimer, 2003). Boys with early work experiences may also become more preoccupied with advancing in the workplace and simply see "little to gain" from further participation in subsequent waves.

*Objective intellectual ability. Intellectual ability* (ZABILITY): a construct created by from four intellectual ability tests (discussed earlier). Persons lower in general intelligence are hypothesized to be less able and willing to cooperate in the study. Further, people lower in intelligence will also be more likely to drop out of high school and be more difficult to locate and induce into further participation.

*Social and economic characteristics. Family's socioeconomic level at Wave 1* (SEL1): a composite measure consisting of father's occupational status (Duncan classification), parents' educational attainment, home environment (family possessions, number of books in the home, and number of rooms per person in the home).[5] SEL may be important because aspirations and achievement are often associated with the socioeconomic status of one's family. Thus, as SEL decreases, aspirations and achievement also decrease. The result is that as SEL decreases, motivation to remain in the study may also decrease. Furthermore, researchers generally recognize that as ones's success or achievement increases, one's motivation to cooperate in a study that inquires about external success also increases. Conversely, less successful people tend to shy away from such studies.

*School experiences. Grade point average in 9th-grade* (GPA1): a variable indicating each boy's self-reported GPA in 9th-grade. It is included because boys with lower GPAs may be less academically able or motivated, more likely to drop out of high school, come from lower socioeconomic backgrounds, and be generally less motivated to remain in the study.

*Failed a grade before 9th grade* (FLUNK1): a dichotomous variable indicating whether or not a boy had ever been held back a grade before entering high school. It is hypothesized to be important because boys who have experienced such an early and visible sign of personal failure may, besides possibly being less able or motivated students, be more highly motivated to seek their rewards outside the school setting. Thus, some of the reasons way GPA is thought to be associated with attrition may also apply to grade failure. Furthermore, because the early years of the YIT study were closely associated with school (as the place where the boys were selected for the study and where they were interviewed) there may be less desire to continue in the study. Furthermore, boys who have failed a grade may be less willing to want to answer questions about their school activities and achievement, two areas closely examined in the study.

### Conceptual Introduction to the Formal Selectivity Bias Problem

This subsection provides a conceptual introduction and review of the formal selection bias problem possibly operating in the YIT data set. The next subsection develops the formal statistical and mathematical foundation of the problem and its solution. The key to understanding selection bias is to be aware of how one's substantive equation may be affected by differential subject retention (or exclusion) and to also be cognizant of how the substantive equation and the so-called selection equation are related to each other. Since a substantive equation marked by selection bias will have a nonzero error term because of the omission of a "selection variable" (which may result in biased parameter estimates and spurious associations), my central goal is to use probit to generate a variable which estimates differential subject retention. The omitted variable problem may be rectified by inserting the "selection variable" (i.e., hazard rate, Mill's ratio) into the substantive equation as another regressor.

Understanding how the probit model and the hazard rates are estimated and applied involves several steps. I begin with equations 1 and 2,

below, and make the assumption that the data contain (i) random observations. Equation 1 is the "substantive equation" of primary theoretical
interest.

(1) $Y_i = \sum b_k X_{ik} + e_i$  ◄—— Substantive equation
   $Y_i$ = Self-esteem at Time 5
   $X_{ik}$ = a vector of k exogenous ("substantive") variables
   $b_k$ = a vector of regression coefficients for k variables
   $e_i$ = an error or disturbance term

If there were no sample selection bias then equation 1 could stand
as it is and be estimated through ordinary least squares because, as I will
explain later, we assume that the error term ($e_i$) equals zero. Furthermore, if
sample bias was suspected and if we had observed a variable that accurately
predicts sample retention (or, conversely, attrition), then we would simply
insert that variable into equation 1 as another exogenous variable, thereby
correcting for sample selection bias by actually accounting for differential
retention of the Wave 5 sample. Since we do not have an observed variable
that accounts for retention, and because it is highly unlikely that a single
source of retention exists, we have what econometricians refer to as an
omitted variable problem (recall the discussion of omitted variables at
the beginning of this section). The hazard rate is, in essence, a proxy for
the omitted variable(s) that might have predicted retention if it had been
observed and included in the substantive equation.

Equation 2, is the "selection equation" and is concerned solely with
predicting differential sample retention (i.e., sample bias or the probability
of remaining in the sample by Wave 5).

(2) $Y_i^* = \sum b_k X_{ik} - u_i$  ◄—— Selection equation
   $Y_i^*$ = an unobserved index representing the probability of being in the
         Wave 5 sample
   $X_{ik}$ = a vector of k exogenous ("selection") variables
   $b_k$ = a vector or regression coefficients for k variables
   $u_i$ = an error or disturbance term

Once estimated, the hazard rates, which are modeled in the selection
equation, can be added to the substantive equation in order to increase the
internal consistency of the predictions by eliminating nonrandom sources
of error in the disturbance term in the substantive equation which stems
from sample bias due variables previously omitted from the substantive
equation (see equation 1.1).

(1.1) $E(\text{self-esteem}|\text{stayers}) = a_0^s + \sum b_{1i}^s X_{i1} + [\text{selection term}] + e_i{}^*.$
$a_0{}^s = $ intercept for the retained sample (i.e., the stayers ["s"])
$b_{1i}^s X_{i1} = $ stayers' regression coefficients and exogenous variables

If there is no selection bias—that is, for example, if there is no omitted variable in the substantive equation which influences both Time 3 self-esteem (or any of the other exogenous variables in the substantive equation) *and* the dependent variable Time 5 self-esteem, then there is no correlated error between the exogenous variables and the endogenous variable. If the error terms are not correlated, then sample selection bias will not be a problem so substantive equation 2 can be solved as it stands. If there is correlated error (between the exogenous variables and the dependent variable) and therefore sample selection bias exists, then including the hazard rate in the substantive equation will produce an unbiased or true estimate of the dependent variable because the corrected error term, called $e_i{}^*$ (in equation 1.1), will have been shorn of error due to sample selection bias. More will be said about this later.

The magnitude of the hazard rate associated with any given subject can be interpreted as the probability of that person being retained in the sample. Berk (1983, p. 391) summarizes the usefulness of the hazard rate this way:

> [T]he hazard rate captures the expected values of the disturbances in the substantive equation after the nonrandom selection has occurred. It was precisely these expected values that are the source of the biased estimates. By including the hazard rate as an additional variable, one is necessarily controlling for these nonzero expectations.

The hazard rate, recall, accounts for the observed and *unobserved* sources of bias due to sample selection. This point will be discussed in detail later.

To recap, the important thing to keep in mind is that if people with particular self-esteem thresholds are more (or less) likely to be retained in the sample, then there is a real possibility that $e_i$ in the substantive equation is nonzero, which raises the specter of correlated error among the variables in the substantive equation. Furthermore, if the disturbances are correlated (because of sample bias) then the sample of Wave 5 "survivors" may be truncated. In this case, one can only observe values of the dependent variable in substantive equation 1 if the endogenous variable in selection equation 2 (i.e., $Y_i{}^*$) is either higher or lower than some threshold. Greene (2002, p. 637) states the problem this way (I have replaced what he calls $z^*$ for what I call $Y_i{}^*$):

[$Y_i^*$] is such that an observation is drawn from [equation 1] only when [$Y_i^*$] crosses some threshold. If the observed data are treated as having been randomly sampled from [equation 1] instead of the subpopulation of [equation 2] associated with the 'selected' values of [$Y_i^*$], potentially serious biases result. The general solution to the selectivity problem relies upon an auxiliary model of the process generating [$Y_i^*$].

The results from estimating $Y_i^*$ in equation 2 are then incorporated in the estimation of equation 1, or really equation 1.1.

A concrete example will help illustrate this twofold estimation and correction process. Suppose that there was something that caused persons with the lowest self-esteem scores to be more likely to drop out of the YIT study. Suppose also that this dropping out factor was causally related to both an exogenous variable in the substantive equation (say, context choice) and to the dependent variable Time 5 self-esteem. If this was true, then the relationship between context choice and later self-esteem would be spurious because there is an *omitted* variable not controlled for in the relationship. Conceptually, the problem could again be graphed as:

where Z = some "dropping out" factor
     X = the exogenous social context variable
     Y = the dependent variable Time 5 self-esteem

Furthermore, if people with lower self-esteem were more likely to drop out of the study (or, conversely, if those with higher self-esteem were more likely to stay), then the sample consisting of the remaining cases (the "usable sample") would be biased because the distribution of the Time 5 self-esteem variable in the sample would be truncated and therefore it would no longer be distributed as it is in the population. Even more importantly, if this occurred the substantive issues and the selection issues become inextricably confounded: All self-esteem estimates would reflect the probabilities of being represented in the sample.

Attention is now shifted to estimating a probit model of sample retention and to calculating what is variously called a selection term, hazard rate, Heckman correction, or inverse Mill's ratio. After that, a section discussing how selection bias actually operates and how it can be detected will presented.

## *Evaluating the Selection Model*

In probit and logit analysis, the significance of a model is tested by the likelihood ratio principle, just as the overall F statistic is used to test the significance of a linear regression model (Aldrich and Nelson, 1984, p. 55). The likelihood ratio statistic can also be thought of as a test of how well the probit model fits the observed data. The likelihood ratio, which follows a chi-square distribution, is computed from equation 3, below (Aldrich and Nelson, 1984, p. 55):

(3) $c = -2 (\log L0 - \log L1)$ where

    $c$ = likelihood ratio statistic (which approximates a chi-square distribution with $k - 1$ degrees of freedom)

    $L0$ = the maximum value of the likelihood function for the null hypothesis when all of the coefficients except the intercept are zero

    $L1$ = the maximum value of the likelihood function for the full model (i.e., the alternative hypothesis) as it is fitted

If the resulting chi-square value of the likelihood ratio is significant, then the null hypothesis that all of the parameter estimates equal zero is rejected. Table 2.8, below, shows the goodness of fit test for the probit model of differential subject selection.

**Table 2.8. Model-to-Data Goodness of Fit Assessment**

| | |
|---|---|
| Log-likelihood Full Model (L1) | $-1057.2$ |
| Log-likelihood Restricted Model Slopes = 0 (L0) | $-1099.0$ |
| Chi-Square Value for Likelihood Ratio (7 d.f.) | 83.654 |
| Significance Level | $p < .00001$ |

Since the chi-square value in Table 2.8, is highly significant ($p < .00001$), the null hypothesis is rejected: The fitted probit model upon which the selection bias term is based is accepted. Or, we can assume that the probit model fits the observed data.

Assessing the explanatory power of the model, however, is less straightforward than assessing its fit through the likelihood ratio (Borooah, 2002). There is no universally accepted measure for summarizing the explanatory power of a probit (or logit) model. However, a pseudo-$R^2$ can be used with caution; it merely suggests the general explanatory power of a probit or logit model because it's upper (1) and lower (0) bounds have no natural interpretation even though the pseudo-$R^2$ probably increases with improvement in model fit (Borooah, 2002, p. 20). In a recent publication, Aldrich and Nelson (1986,

p. 143) give a stern warning about putting too much stock in the pseudo-$R^2$. Instead, they believe greater use of t and chi-square statistics as goodness-of-fit measures.

Aldrich and Nelson's (1984, p. 57) pseudo-$R^2$ (i.e., Pearson's C) is:

(4) pseudo-$R^2$ = c/(N + c) where,

    c = the chi-square statistic for the overall model fit (see equation 3, above)

    N = total sample size (2,070)

In my probit model, the pseudo-$R^2$ = 83.654/(83.654 + 2070) = .039, which indicates that the explanatory power of the selection model is weak.

The real test of the usefulness of the selection model—and thus the selection term or hazard rate—is to actually compare the $R^2$s and b-coefficients in the substantive model when the selection term is included and when it is not. If the selection term significantly increases the amount of explained variance, or if the b-coefficients for the explanatory variables change significantly, then there is good evidence for including the hazard rate in the substantive model. Assessment of the actual usefulness of the selection term is performed in chapter 4.

*Interpreting the b-Coefficients in the Probit Model*

The magnitude of the maximum likelihood b-coefficients reported in Table 2.9, below, do not have straightforward interpretations and cannot be directly compared with each other. This is true because $P(Y = 1)$ is a function of Z (where $Z = \sum b_k X_k$ = the stayer potential index) and Z is nonlinear and therefore characterized by a nonconstant rate of change

**Table 2.9. Probit (Selection) Equation for Remaining in the Wave 5 Sample**

| Variable | Maximum likelihood of $B_k$ | Standard error of $B_k$ | T-Ratio | Sig. |
|---|---|---|---|---|
| Intellectual ability | −.0242 | .0100 | 2.405 | .016 |
| Socioeconomic level | .0368 | .0453 | 0.811 | .417 |
| Parents divorced | −.3785 | .1085 | −3.488 | .000 |
| Number of parents | .1963 | .1091 | 1.800 | .072 |
| Urbanicity | −.1601 | .0632 | −2.534 | .011 |
| Failed a grade | −.2710 | .0798 | −3.393 | .001 |
| Ninth grade GPA | .0114 | .0050 | 2.251 | .024 |
| Worked In tenth grade | .0326 | .0656 | 0.497 | .619 |
| Intercept | −.8820 | .3734 | −0.236 | .813 |

(see also Aldrich and Nelson, 1984, pp. 41–44; Hanushek and Jackson, 1977, pp. 204–207). Instead, the key to interpreting the b-coefficients is to note their sign and significance. The sign of a b-coefficient is interpreted similarly to a linear regression coefficient. For example, a positive $b_k$ indicates that the probability of staying in the sample (i.e., $P[Y = 1]$) increases as the value of the exogenous variable ($X_k$) increases. A negative $b_k$, on the other hand, indicates that the probability of being in the Wave 1 sample decreases as $X_k$ increases.

In contrast to Bachman et al., socioeconomic level did not significantly predict staying in the sample (or, converse, attriting). A boy was more likely to be present in the Wave 5 sample if both of his parents were alive during the Wave 1 interview ($b = .196$). Academic success was also positively related to staying in the sample, so that as GPA increased, so did the probability of being in the Wave 5 sample ($b = .011$). However, a boy had a higher probability of dropping out of the Wave 5 sample if he was from a broken home in 10th-grade (as measured by parental divorce), if he was raised in a rural area, and if he had ever failed a grade before the Wave 1 interview. The b-coefficients associated with these last three variables are $-.378$, $-.16$, and $-.271$, respectively.


## An Exploration of Global Self-Esteem's General Dimensions

### *Background*

Each of us possesses a strong propensity to cast ourselves in the best possible light, to accentuate the positive. Yet it is precisely this emphasis which has distracted social scientists from the critical, disparaging aspects of the self, our self-perceived negative side, and its consequences for well-being. I look at these issues by comparing two major positions regarding the nature of global or general self-esteem. The first position views global self-esteem as a unidimensional phenomenon best reflected by incorporating positive and negative self-evaluations in a summary measure (e.g., Rosenberg 1965, 1979; Rosenberg, Schooler, and Schoenbach, 1989; Rosenberg, Schooler, Schoenbach, and Rosenberg, 1990; Carmines and Zeller, 1979; Fleming and Courtney, 1984). Thus conceived, global self-esteem is a type of general self-esteem defined as a "positive or negative attitude toward a particular object, namely, the self" (Rosenberg, 1965, p. 30). When global self-esteem is high, one has self-respect and a feeling of worthiness, and yet acknowledges faults and shortcomings. When it is low, people lack self-respect and see only their weaknesses and thus consider themselves "unworthy, inadequate or otherwise seriously deficient" (Rosenberg, 1979, p. 54).

The other position sees within global self-esteem positive and negative self-evaluative components (e.g., Kaplan and Pokorny, 1969; Kaplan, 1971; Kohn, 1977; Kohn and Schooler, 1983; Goldsmith, 1986; Owens, 1992, 1993):[6] Here self-esteem is comprised of general self-denigrating and general self-affirming subscales, or critical, self-deprecation and positive, self-worth components (Kohn, 1977; Kohn and Schooler, 1983). This division also neutralizes criticism that researchers think solely in terms of global or fixed levels of self-esteem, a view that may militate against its more precise examination (Gergen, 1971; Harter, 1985).

This section examines theoretical issues concerning negative and positive attitudes toward the self. Dimensionality is tested through exploratory and confirmatory factor analyses. I go beyond previous studies by attending to the theoretical reasons underpinning bidimensional self-esteem, by explicitly comparing the structure of global self-esteem with those of general self-deprecation and general self-worth, and by testing the structural invariance of each construct from late adolescence to early adulthood. Estimation of a second-order measurement model and examination of the differential association of self-deprecation and self-worth with measures of socioemotional are also presented.

*Negative Evaluations of the Self*

Self-concept theory credits two central motives for the protection and maintenance of one's self-picture: self-esteem and self-consistency (see Rosenberg, 1979). The former induces individuals to think well of themselves (see e.g., Kaplan, 1975; Allport, 1961; Rosenberg, 1979). Indeed, many self theorists regard this motive as universally dominant (see Kaplan, 1975). The self-consistency motive (Lecky, 1945) asserts that people struggle to validate their self-images, even when their images and evaluations are negative. Here people are rather self-centered and insular, such that ideas at variance with their value-system and self-view are rejected, unless a cognitive reorganization occurs (Lecky, 1945).

How can these two superficially incompatible motives be reconciled with my assertion that negative self-evaluations merit their own theoretical and empirical examination? Through a brief overview of self-verification theory and theories based on positive strivings, I suggest that individuals may sometimes perceive benefits from negative self-images and evaluations.

Drawing on cognitive psychology and symbolic interactionism, self-verification theory poses as a corrective to self-consistency theory's insistence that a consistent self-concept is an end in itself (Swann, Stein-Seroussi, and Giesler, 1992). In this view, negative self-conceptions may help maintain a viable self-system and predictable, orderly social relations. People understand that a stable self-concept promotes successful negotiation of social reality by yielding interpersonal predictability, thus preferring reflected appraisals that confirm their self-images over those that do not, even if negative (see also Taylor and Brown, 1988). Accordingly, perceptions of prediction and control may be bolstered by confirming one's (negative) self-conceptions and self-evaluations (see also Pittman and Heller, 1987; Gecas and Schwalbe, 1983; Greenwald, 1980).[7] This means successful social relations hinge on recognizing how others perceive you (Cooley, 1922; Mead, 1934). Thus, a perceived disjuncture between one's self-concept and feedback from others may arouse fear of disharmonious interpersonal relations and misunderstandings.

Theories based on positivity strivings acknowledge that while everyone may want positive reflected appraisals (cf. the self-esteem motive), people who offer negative appraisals to people making negative self-judgments, may actually satisfy the target's positivity strivings. This seeming contradiction may be reconciled four ways (Swann et al., 1992, p. 393): (1) People with poor self-images may seek negative feedback from others in order to identify and rectify problem behaviors. (2) As self-attribution theory suggests (Kelley, 1971), people feeling worthless may also seek critical associates in the hope that winning them over may prove their worth after all. (3) If people prefer associating with those who share their values and beliefs as balance theory suggests (Heider, 1958; Rosenberg and Abelson, 1960), then people with negative self-concepts may try to validate their negative self-attitudes by choosing associates who appraise them unfavorably. (4) People with negative self-images may seek relationships with those who give them negative feedback in the gratifying belief that they have intelligent and perceptive associates.

Viewed in the above light, theories of self-esteem, self-verification, and positive strivings point to the merit in isolating and examining content-free, general negative self-evaluations and images, or, in the present context, self-deprecation. Indeed, the theories would seem to expect both poorer psychological well-being and more realistic self-orientations among self-denigrators than those holding positive attitudes of worth (see Taylor and Brown, 1988 for a review of negative realism). (See Rosenberg and Owens [2001] for an extensive review of low self-esteem people.)

*Positive Evaluations of the Self*

As a basic human motivation, self-efficacy has been hypothesized to constitute an "effectance motivation" that compels judgment of one's own competence and efficacy versus ineptitude and ineffectiveness (Gecas and Schwalbe, 1983). By extension, the development of self-worth should be associated with an awareness of or concern for self-efficacy, while coextensive development of self-deprecation should follow from an awareness of personal actions self-rated as inept, ineffectual, or unsuccessful. The effectance motivation should also impel more focus on one's own varying degrees of ability, competence, and efficacy—all self-worth attributes—than on the inefficacy that fosters self-deprecation (Greenwald, 1980). Moreover, self theorists tend to agree that a viable and effective self-system is served by a strong inclination to positively construe one's self-images and self-evaluations by manipulating or casting self-attributions, social comparisons, reflected appraisals, behaviors, and intentions in the best light possible (see Markus and Wurf, 1987). Several studies show that self-enhancement motivates people to selectively interpret and remember events positively, highlighting successes and modifying recall to support favorable self-concepts and evaluations (see Markus and Wurf, 1987; Greenwald, 1980). This idea comports with self-esteem theory, wherein people are motivated to protect and enhance their self-esteem (Rosenberg, 1979).

To summarize, much research attention has focused on global and positive self-esteem, but little on general self-deprecation. Indeed, positive self-esteem (comprised solely of enhancing self-statements) is commonly referred to simply as self-esteem (e.g., O'Malley and Bachman, 1983; Kanouse, Haggstrom, Blaschke, Kahan, Lisowski, and Morrison, 1980), evincing the short shrift given the negative dimension of global self-esteem. Yet just as negative self-conceptions sometimes catalyze change in one's self-system (Markus and Wurf, 1987), negative self-evaluations, though discomforting and difficult for the individual to acknowledge, may serve a parallel function. Beck (1967), for example, implicates negative self-evaluations with increased depression, which accords well with cognitive consistency theory (see also Rosenberg et al., 1989) and may presage personal growth. Again, much work remains in our exegesis of self-deprecation, as the preponderance of thought and energy—by individuals and social scientists alike—has sought to explicate the positive dimension of self-esteem at the expense of better understanding the course and effect of its negative dimension.

## Analysis Strategy

Self-esteem's dimensionality is examined through confirmatory factor analyses with support from exploratory factor analysis and correlations. To avoid capitalizing on chance, the exploratory analysis was performed on one random half of the sample, the confirmatory analyses on the other. The former analysis takes the conventional route in ascertaining whether a bidimensional self-esteem solution obtains and warrants further, more conclusive, investigation. Here the self-esteem items are analyzed via direct oblimin oblique rotation (the factors are assumed to be correlated). The confirmatory analyses explicitly test the dimensionality of the constructs. The first model specifies a unidimensional self-esteem construct, measured by positive and negative items; the second posits two latent general self-esteem dimensions corresponding to self-deprecation and self-worth, respectively; the third model assesses the relation of the two dimensions to the general concept of self-esteem through a second-order factor analysis.

## Results

### Exploratory Findings

In the oblique factor analysis, the pattern matrix (path coefficients) and structure matrix (correlations between the factors and the indicators) indicate two rather distinct, unambiguous, factors—self-worth and self-deprecation—are emergent (Table 2.10).[8]

### Confirmatory Findings

Confirmatory factor analyses explore dimensionality explicitly and more conclusively, assessing whether a one-factor model, with all self-esteem items forced to load on a single construct, fits the data better than a two-factor model, with positive and negative self-evaluations loaded on

## Table 2.10. Oblique-Factor Solution (Oblimin Rotation) of Self-Esteem Variables

| | Wave 3 (Age 18) (N = 747) | | | | Wave 4 (Age 19) (N = 665) | | | | Wave 5 (Age 23) (N = 697) | | | |
| --- | --- | --- | --- | --- | --- | --- | --- | --- | --- | --- | --- | --- |
| | Pattern Matrix | | Structure Matrix | | Pattern Matrix | | Structure Matrix | | Pattern Matrix | | Structure Matrix | |
| | F1 SW | F2 SD | F1 SW | F2 SD | F1 SW | F2 SD | F1 SW | F2 SD | F1 SW | F2 SD | F1 SW | F2 SD |
| Positive self-esteem items | | | | | | | | | | | | |
| I'm of worth | .691 | −.007 | .694 | −.281 | .743 | .006 | .741 | −.306 | .740 | .009 | .736 | −.296 |
| I Have good qualities | .802 | .038 | .787 | −.280 | .737 | −.063 | .764 | −.374 | .807 | .045 | .789 | −.287 |
| I do things well | .669 | −.066 | .696 | −.332 | .752 | .013 | .752 | −.305 | .786 | .103 | .743 | −.220 |
| I Have a positive self | .518 | −.290 | .634 | −.496 | .559 | −.297 | .683 | −.532 | .521 | −.284 | .639 | −.499 |
| I'm useful | .702 | .072 | .673 | −.206 | .737 | .141 | .677 | −.168 | .642 | −.071 | .672 | −.336 |
| I do a job well | .593 | .026 | .582 | −.209 | .538 | −.073 | .569 | −.300 | .563 | −.015 | .570 | −.248 |
| Negative self-esteem items | | | | | | | | | | | | |
| I'm Not Proud | −.038 | .688 | −.312 | .703 | .115 | .564 | −.353 | .613 | .064 | .655 | −.205 | .628 |
| I'm No Good | .047 | .831 | −.282 | .812 | .015 | .827 | −.363 | .833 | .041 | .771 | −.359 | .788 |
| I can't do right | −.047 | .668 | −.312 | .687 | .075 | .773 | −.249 | .741 | .014 | .708 | −.306 | .714 |
| My life's not useful | .036 | .760 | −.265 | .745 | .001 | .793 | −.332 | .793 | .048 | .746 | −.356 | .766 |
| Eigenvalue | 3.62 | 1.43 | | | 3.92 | 1.4 | | | 3.73 | 1.37 | | |
| Factor correlation (F1, F2) | −.397 | | | | −.420 | | | | −.413 | | | |

*Note:* See Table 2.5 for wording of self-esteem items. SW = self-worth, SD = self-deprecation.

separate constructs. LISREL 7 (Joreskog and Sorbom, 1989) is used to estimate parameters and test each model's overall goodness-of-fit, after imposing theoretically derived constraints on the data. Fit assessment helps determine whether a given theoretical model adequately describes the pattern of relations within a set of data. I report four fit measures: (1) chi-square/degrees-of-freedom ratio (fit ratio), (2) goodness of fit index (GFI), (3) normed fit index (NFI), and (4) root mean square residual (RMR).[9]

To keep the models as general as possible, I exclude correlated error terms, except where indicated. Inspection of the modification indices supports this decision: no significant improvement in fit would result by including them. Furthermore, no theoretical reason compels correlating error terms in the present models, especially since I am generally not making longitudinal estimates (cross-lagged error terms are specified in the full substantive model in chapter 4).

The unidimensional, one-factor model (Figure 2.4) fits the data extremely poorly. For Wave 3 (age 18) the fit ratio is 10.5, the GFI .907, and the NFI .827. The Wave 4 (age 19) model fits the data more poorly—the fit ratio is 11.7, the GFI .878, and the NFI .781. Wave 5's model also fits very poorly: the fit ratio is 15.4, the GFI .845, and the NFI .790.

The bidimensional model fits the data much better (Figure 2.5). Here the Wave 3 fit ratio is 4.5, the GFI a respectable .965, and the NFI a relatively large .927—a combination that indicates a good alternative fit. Waves 4 and 5 present similar pictures. Wave 4's fit ratio is 3.5, the GFI 0.970, and the NFI .922. In Wave 5 the fit ratio is a somewhat higher 5.9, but the GFI quite acceptable at .947 as is the NFI at .926.

Comparing the intrawave models indicates that the bidimensional models, nested within the unidimensional, are significant improvements over their unidimensional counterparts (see Table 2.11). Taken overall, strong empirical evidence supports the bidimensional self-esteem model.

Table 2.11. **Within-Time Fits of Unidimensional and Bidimensional Self-Esteem (SE) Measurement Models**

| Models | Chi-Square | df | Chi-Square | df | Chi-Square | df |
|---|---|---|---|---|---|---|
| | Model A | | Model B | | Model C | |
| Unidimensional | 366.41 | 35 | 409.45 | 35 | 540.15 | 35 |
| Bidimensional | 153.44 | 34 | 118.42 | 34 | 200.75 | 34 |
| Difference | 212.97* | 1 | 291.03* | 1 | 339.40* | 1 |

*$p < .0001$.

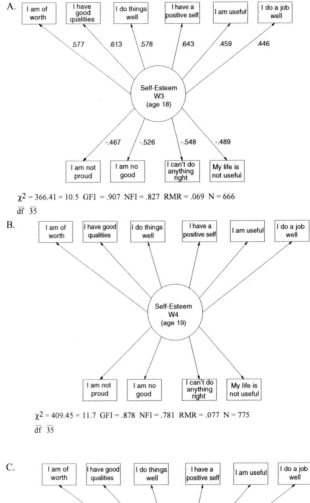

A.

$\dfrac{\chi^2}{\overline{df}} = \dfrac{366.41}{35} = 10.5$  GFI = .907  NFI = .827  RMR = .069  N = 666

B.

$\dfrac{\chi^2}{\overline{df}} = \dfrac{409.45}{35} = 11.7$  GFI = .878  NFI = .781  RMR = .077  N = 775

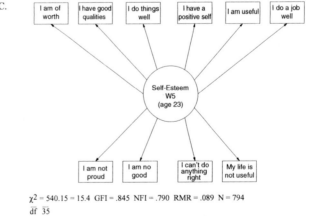

C.

$\dfrac{\chi^2}{\overline{df}} = \dfrac{540.15}{35} = 15.4$  GFI = .845  NFI = .790  RMR = .089  N = 794

**Figure 2.4.** Confirmatory factor analysis of one-factor measurement models of unidimensional self-esteem (standardized solution)

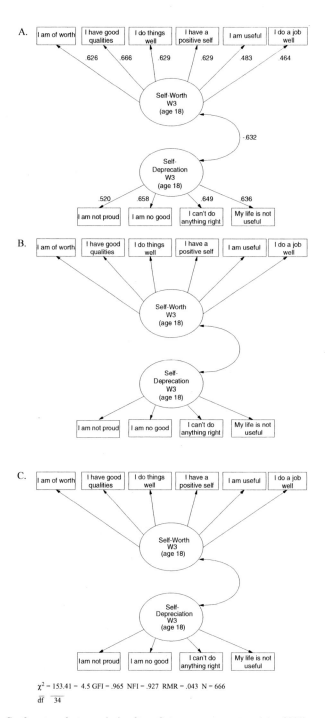

$$\frac{\chi^2}{df} = \frac{153.41}{34} = 4.5 \text{ GFI} = .965 \text{ NFI} = .927 \text{ RMR} = .043 \text{ N} = 666$$

**Figure 2.5.** Confirmatory factor analysis of two-factor measurement models of bidimensional self-esteem (standardized solution)

**Table 2.12. Results of Structural Invariance Tests of
Alternative Self-Esteem Measurement Models (Wave 3
and Wave 5)**

| Models | Chi-Square | df | |
|---|---|---|---|
| Global Self-Esteem | | | |
| 1. Unconstrained | 1554.06 | 382 | |
| 2. Constrained | 1583.88 | 400 | |
| Self-Worth | | | |
| 1. Unconstrained | 361.62 | 120 | |
| 2. Constrained | 373.26 | 130 | |
| Self-Deprecation | | | |
| 1. Unconstrained | 103.90 | 43 | |
| 2. Constrained | 110.65 | 49 | |
| Model Comparisons | Chi-Square Difference | df Difference | Prob |
| 1 and 2 | 29.82 | 18 | $p < .05$ |
| 3 and 4 | 11.64 | 10 | ns |
| 5 and 6 | 6.75 | 6 | ns |

*Note:* All models include cross-lagged correlated error terms across all waves.

Structural invariance was tested by comparing each model's fit with the observed data when their corresponding unstandardized lambda coefficients (i.e., those parameters expressing the relationship between the indicators and the latent constructs) were allowed to be freely estimated versus constrained to be equal across waves. (Structural invariance exists when a construct is characterized by the same dimensions across waves and when the pattern of relations among its measured attributes persists over time.) Since the unique variances of the indicator errors for the same items were expected to covary across waves, I allowed their error terms to be correlated across time. Table 2.12 results indicate that global self-esteem's measurement structure is not stable (i.e., the constrained model's fit is significantly poorer than the unconstrained, $p < .05$). However, the structures of self-deprecation and self-worth are invariant (i.e., each constrained model's fit is not significantly poorer than the respective unconstrained).[10]

The second-order confirmatory factor analysis tests a hierarchical model in which a more general and abstract latent variable (the second-order global self-esteem factor) directly influences two "less" abstract first-order latent variables (self-deprecation and self-worth), which in turn directly affect the observed self-esteem indicators (Figure 2.6).[11] As expected, self-esteem strongly affects the first-order factors, with a somewhat stronger

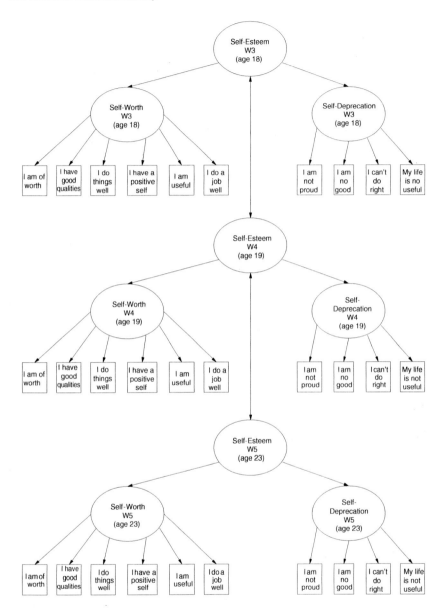

$$\frac{\chi^2}{df} = \frac{1378.17}{397} = 3.5 \quad \text{GFI} = .87 \quad \text{NFI} = .87 \quad \text{RMR} = .094 \quad \text{N} = 647$$

**Figure 2.6.** Second-order confirmatory factor analysis of bidimensional self-esteem (standardized solution)

influence on self-worth than self-deprecation. The model fits the data reasonably well. The fit ratio is 3.07, the RMR .049, the GFI .88, and the NFI .87. (Identification precludes incorporating correlated error terms.) The second-order factor analysis lends additional support for a bidimensional view of self-esteem.

Construct validation suggests that self-deprecation (SD) and self-worth (SW) may be distinguished by their differential association with measures of socioemotional well-being linked with self-esteem (e.g., Rosenberg, 1979, 1985; Cherlin and Reeder, 1975). Self-deprecation appears significantly related to "worrying about what others think of me" (.22), self-blame (.17), an index of trouble with parents (.18), and an index of lack of trust in others ($-.09$), yet none are significantly associated with self-worth. (See Bachman et al., 1978 for index descriptions.) Self-deprecation is also more highly associated than self-worth with indexes of emotional dependence (SD = .24, SW = .09), negative affect (SD = .69, SW = $-.30$), depression (SD = .62, SW = $-.32$), anxiety (SD = .50, SW = $-.17$), anomie (SD = .66, SW = $-.25$), and resentfulness (SD = .56, SW = $-.23$). Unlike self-deprecation, self-worth is significantly associated with indexes of valuing self-control (.16) and kindness toward others (.15), as well as one of self-utilization desires (.27). Interestingly, self-worth is significantly related to an aspect of guilt (being punished by one's conscience) (.14), while being unrelated to self-deprecation. The overall pattern suggests that self-worth is associated with issues of self-development and prosocial behavior, self-deprecation more with indicators of diminished psychological and social well-being, particularly distress, isolation, and dependence.[12] (See Rosenberg and Owens [2001] for a detailed examination of low self-esteem).

## *Conclusion*

This section finds strong theoretical, methodological, and substantive support for a bidimensional self-esteem measure composed of a general self-worth subscale and an often overlooked general self-deprecation subscale. Confirmatory factor analyses support such a construct, and not a unidimensional one. The second-order factor analysis is consistent with this finding, and also shows self-worth has a somewhat stronger association with global self-esteem than does self-deprecation. Although each model leaves some unexplained variance around the measured variables, good reason favors both the bidimensional and second-order models as stronger specifications of the theory. Furthermore, the results indicate that self-deprecation and self-worth are structurally invariant between adolescence and adulthood, while global self-esteem is not, which raises troublesome questions about the latter's use in change analysis (see Bengtson, Reedy, and Gordon, 1985). Finally, construct validation shows the bidimensional construct discriminates among prosocial and well-being measures. In light of my initial theoretical arguments, these findings support a bidimensional approach.

Self-esteem's usefulness in research has been weakened by uncritical employment of unidimensional or bidimensional self-esteem constructs. This obscures important theoretical, empirical, and substantive nuances. Bidimensional self-esteem models, for example, most frequently explore self-worth. Although its use is beginning to increase (Owens, 1993; Mortimer, Finch, Shanahan, and Ryu, 1992), the negative dimension lacks but merits equal attention (Schwalbe and Staples, 1991). Owens' (1993) analysis of the reciprocal effects of self-deprecation and self-worth on depressed mood, for example, shows that self-deprecation's impact on depressive affect ($B = .51$, $p < .001$) is double self-worth's ($B = -.21$, $p < .001$), while the reciprocal impact of depressive affect on self-deprecation ($B = .69$, $p < .001$) is twice that on self-worth ($B = .34$, $p < .001$). Mortimer et al. (1992) report that work stress and the likelihood of early high school employment increase adolescent boys' self-deprecation ($B = .294$, $p < .001$ and $B = .186$, $p < .01$, respectively), but neither significantly relates to positive self-esteem.

The construct validity analyses show that self-deprecation is significantly or more highly associated than self-worth with variables describing people at variance with themselves or others. Consistent with this, psychoanalytic theory recognizes that persons lowest in self-esteem experience the most alienation and conflict between their real self and the perceived selves associated with their various roles (Horney, 1950). Rosenberg and Owens (2001) also emphasize low self-esteem's relation to psychological and emotional distress (e.g., depression, anxiety, vulnerability, negative affect states). Clearly, more use of general self-deprecation constructs would deepen insight into self-esteem.

Future research might examine why people focus on particular aspects of self-evaluation—either positive or negative. The relationship of self-worth and self-deprecation to guilt and shame is illustrative. My analyses suggest that increased feelings of guilt are positively associated with self-worth (and global self-esteem), but are unrelated to self-deprecation. Why? Guilt is a powerful emotion predicated on fear of punishment arising from violating accepted norms (see Kemper, 1987; Scheff, 1987). Pangs of conscience or guilt may direct people's attention to the wrong done another (Kemper, 1987), thus steering their social participation in a more constructive, rewarding manner (see Wallace, 1988) or evoking healthy prosocial behavior leading to a reconciliation with self or others (see Rosenhan, Salovey, Karylowski, and Hargis, 1981; Harris, Benson, and Hall, 1975). Conversely, self-blame may be associated with self-deprecation because it is expressed as inward punishment and self-reproach, which in essence focuses shame on the self and highlights its unworthiness, weakness, and other negative features (see Kemper, 1987; Scheff, 1987). A bidimensional view of self-esteem reveals these nuances and subtleties.

This examination of Rosenberg-type global self-esteem illuminates the dimensionality of this widely used family of global or general self-esteem constructs, and especially on the rather neglected dimension of general self-deprecation. The construct validity analyses highlight the need for understanding better the differential roles shame and guilt play in self-deprecation and self-worth. Although Owens and King (2001) have initiated an examination of the role of gender, race, and other age-groups in the usefulness of bidimensional general self-esteem and the maintenance of well-being, additional research is needed. As others have persuasively argued (e.g., Wylie, 1974; Crandall, 1973), we do not so much need new self-esteem measures as we need better understanding—and more discriminating use—of those already available.

## Summary and Findings

This chapter focused on five methods-related tasks that were discussed in five corresponding sections. The tasks were to: (1) describe the Youth in Transition data set, (2) define the post-high school social contexts of theoretical interest, (3) specify and analyze a procedure for correcting for sample selection bias, (4) measure the major independent variable constructs, and (5) present a measurement model of the dependent variables. The first section showed that the Youth in Transition data set is well suited to the key theoretical interest of this book: examining the effect of post-high school social context on psychological development as it pertains to changes in self-esteem. The data set consists of five waves (1966,

1968–1970, 1974) that were gathered from a nationally representative sample of 10th-grade boys enrolled in an American public high school in 1966. After excluding racial minorities from my research sample, because of a possible "school effect" in the minority sample due to the selection of blacks from relatively few mostly all-black schools, the book begins with a Wave 1 sample of 2,177 whites.

The social contexts (i.e., work, military, and college) were defined in terms of two key requirements: mutual exclusiveness and a major post-high school involvement in one of the contexts. Mutual exclusivity means that categorization in one context excludes categorization in another. The "major involvement" condition was derived from both the quantity and the overall quality of the context experience. The quantity issue centered on a boy's time-in-context, while the quality issue focused mostly on the nature of the socialization demands placed upon the entrant. The work group consists of subjects who worked mostly full-time after high school, did not serve in any branch of the military (including the guard or reserves), and received no more than one year of post-secondary schooling. The college group consists of everyone who was mostly a full-time student in four out of the five years following high school and never served in any branch of the military. The military group, finally, is comprised of those who served on active Federal military duty in the Army, Navy, Air Force, Marines or Coast Guard for at least 18 months (the minimum full-term of service).

A confirmatory factor analysis of the Wave 3 (12th-grade, 1969) and Wave 5 (1974) self-esteem indicators showed that the data are most adequately characterized by a two-factor structure consisting of self-worth (i.e., positive self-esteem) and self-deprecation (i.e., negative self-esteem). The possibility of response set bias was examined and argued against. All substantive analyses are focused on self-worth.

The two important independent variable constructs measured were family of origin socioeconomic status and intellectual ability. Because of large amounts of missing or possibly unreliable data, the SES construct consists of parents' education measured in Wave 1 and family income measured in wave 3. The intellectual ability construct was composed of four tests: a general test of intelligence, reading comprehension, vocabulary, and arithmatic reasoning. A reliability analysis indicated that the construct achieved adequate reliability (standardized alpha equals .82).

Finally, a major section dealing with sample selection bias was presented. Although selection bias may have many sources, it was argued that the most probable form of bias in the longitudinal YIT data set would stem from attrition-based bias. Very briefly, selection bias from attrition is present when a previously random sample becomes nonrandom because subjects were systematically included (or excluded) from subsequent samples. Without correcting for the possibility of sample bias, one's parameter estimates will be biased and one's findings may be spurious since one will essentially have an omitted variable problem if differential sample selection is not accounted for. In order to correct for attrition from the Wave 5 sample, a complex probit model of the factors hypothesized to influence attrition was specified and tested. The real test of selection bias is assessed in the Chapter 4 when the selection terms is included as another regressor in the substantive equation: if the selection term is significant, then attrition bias is operating; if the term is not significant, then bias probably does not occur.

## Notes

1. The time served in the military was restricted to those who served 18 months or more on active duty, except for those who reported having served in Vietnam, for whom no time in service restriction was made. Even though the minimum time of enlistment or conscription was two years in the Army and Marines (the Air Force and Coast Guard had four year enlistment minimums) special early release policies often made it possible for active duty military personnel to obtain four to six month early releases from active duty. Beginning in the fall of 1971 and extending through 1972, for example, the Army released several hundred thousand draftees within six months of their two-year service commitment under a forced early release policy. This policy was initiated when the army's troop commitment in Vietnam was significantly reduced and it was attempting to build an all volunteer army. Other branches of the military, along with the army, also had early release programs for persons planning to continue their educations after leaving the service. As such, the 18 month minimum military duty commitment accurately reflects a full term of military duty for many veterans of this era.

   Furthermore, even though I do not have data on the type of discharge the veterans received, it is probably safe to assume that the men who served less than 18 months were discharged early because of behavioral, emotional or physical problems manifested in the early stages of their duty (except for those who may have suffered an injury or disease leading to an early release). As such, most of the subjects who served less than 18 months would probably not typify most veterans and may in fact represent a more deeply troubled minority.

2. While gender differences in level of self-esteem have been observed elsewhere (see Owens and Serpe, 2003; Owens and King, 2001; Rosenberg, 1979; Simmons and Blyth, 1987), gender appears unrelated to self-esteem dimensionality (Owens and King, 2001; Simmons and Blyth, 1987).

3. This differential probability would be driven by the likelihood that a subject would be included (or excluded) simply on the basis of whether or not he took one of the three "traditional" paths to adulthood. This subject exclusion on the basis of social context would be all the more serious if the dependent variable was truncated as a result. Truncation (or attenuation as it is also called) is the situation in which the distribution of scores or values on a dependent value does not reflect the full range of scores in the population. The most typical cases of truncated dependent variables occur when the upper or lower ends of the population distribution are excluded. Maddala (1983, p. 150) and Berk (1983, p. 387) show that when this occurs biased regression estimates are obtained. In my case, it would be a particular problem if the self-esteem scores for the "other" group were generally higher or lower than the scores for the three social context groups. Maddala (1983, Chapter 9, p. 257ff) devotes a whole chapter to the problems that accompany models with self-selectivity.

4. The issue of using strictly exogenous variables in the selection equation remains controversial. On the one hand, there is the argument that including endogenous variables as explanatory variables in the selection equation will lead to biased b-coefficients and hazard rate estimates. But this is a problem common to all causal models, not just sample selection models. On the other hand, there is the argument that including purely exogenous variables in the selection equation is too restrictive because one usually is confined to demographic variables such as sex, race, age, and so forth that have little explanatory power. Those who argue the latter point stress that the most important thing is to avoid variables that are highly related to the dependent variable in the selection equation, and may be considered a function of the dependent variable in the substantive equation.

5. Socioeconomic level (SEL) was chosen as a selection equation variable for three reasons. First, it is the socioeconomic status-related variable that is used by Bachman et al. (1970, pp. 9–14) in all of their analyses, including their own analyses of attrition. Using this variable in the analyses would allow comparison with their earlier findings that persons lowest in SEL were more likely to attrite. Second, it offers a good summary measure of family background. Bachman (1970, p. 10) presented the argument this way:

> Our interest is not . . . focused primarily on stratification by status or prestige. We consider that a number of intercorrelated factors in a family— such as parents' educational and occupational attainment, income, and possessions of certain goods (e.g., books, typewriters, cameras)—are all determinants of whether a home is a rich environment for learning, an environment in which education and achievement are likely to be encouraged.

Finally, Bachman (1970) further justifies the use of SEL as a composite family background index because its components are strongly intercorrelated and because "it captures most of the predictive power that the components would have if they were permitted to operate separately" (p. 10). Third, a standard socioeconomic status variable composed of education, occupation and income was not possible because over one-third of the boys did not know their parents' income, and the survey staff never asked the parents directly. And even though they may be a tenuous link, the household possessions may actually serve as an indirect measure of family wealth through conspicuous consumption (Veblen, 1948).

6. Facet specific self-esteem is distinguishable from bidimensional general and unidimensional global self-esteem. The former is situationally dependent or referent specific, such as physical self-esteem or academic self-esteem, and reflects self-esteem in specified arenas or contexts (see Rosenberg, Schooler, Schoenbach, Rosenberg, 1995). Coopersmith's (1967) Self-esteem Inventory with its peers, parents, school, and personal interests subareas is illustrative.

7. Whereas Gecas (1982, 1989), Gecas and Schwalbe (1983), and others (e.g., Bandura, 1977, 1981; Harter, 1985) distinguish self-esteem from self-efficacy, particular Rosenberg-type global self-esteem indicators reflect both self-orientations. For example, Table 2.5 items can be grouped into two general self-esteem classes (see Gecas, 1982): (1) moral worth or self-worth (e.g., having a sense of virtue, justice, reciprocity, and honor) and (2) self-efficacy (e.g., having a sense of competence, power, and human agency). Items 1, 7, and 8, and to a somewhat lesser extent 2 and 4 reflect a sense of self-worth, items 3, 5, 6, 9, and 10 a sense of self-efficacy.

8. An unreported varimax rotated principal components analysis produces the same two highly distinct self-esteem factors, as others have found (e.g., Carmines and Zeller, 1979; Goldsmith, 1986; Kaplan and Pokorny, 1969).

9. The fit ratio generally should not exceed about 3 (Carmines and McIver, 1981) to 5 (Wheaton, Muthen, Alwin, and Summers, 1977). A nonsignificant fit ratio shows that the variance-covariance matrix derived from the theoretical model is similar to that observed, as one would hope. The GFI shows the amount of variance and covariance the model accounts for, and ranges from 0 (no account) to 1 (perfect account). A value above .9 is generally acceptable, although no clear consensus on cut-off level prevails (Joreskog and Sorbom, 1989). The NFI compares a simple, highly restricted baseline model to one's theoretical model (Bentler and Bonett, 1980). Again, while no clearly agreed upon cut-off or referent exists, an NFI greater than about .9 is generally acceptable. Finally, since the RMR is a measure of the square root of the mean of the squared discrepancies between the input variance-covariance matrix and the model's reproduced variance-covariance matrix, a small value near zero indicates a good model fit (Bollen, 1989).

10. Structural invariance exists when a construct is characterized by the same dimensions across waves and when the pattern of relations among its measured attributes persists over time.

11. Since the second-order factor comprises two indicators, identification required longitudinal data (see Heise, 1970). Identification also required fixing the variance of the second-order factors (i.e., the independent variables) at 1 (see Bollen, 1989; Bentler, 1992).

12. Reliance on global self-esteem would have either missed or under-stated the association with all of the well-being measures, but especially with worry about others' opinion of self, emotional dependence, self-blame, and to a somewhat lesser degree, anxiety and anomie.

# 3

## Choice of Paths after High School and the Class of 1969 in Sociohistorical Perspective

The transition to adulthood is a complex process which entails a wide variety of choices, influences, and pressures. Out of the turbulence of adolescence, perhaps one of the most far-reaching decisions a young person will make is what to do after high school. This chapter examines the complex factors which lead a young person to enter the important social contexts of the full-time labor force, the military, or college in the years following high school. Few publications, if any, take a polychotomous perspective on post-high school context choice. Most previous research focuses on single or dichotomous choices—such as choosing whether or not to go to college. As such, a major point stressed throughout this chapter is that post-high school social context choice, like occupational choice generally, is a multidimensional phenomenon. Since the choice of a social context is in many respects preliminary to selecting a particular occupation, the occupational choice literature is supplemented with insights from the large achievement and ambition literature (see Spenner & Featherman, 1978) and from empirical studies dealing with context choice.[1]

The three post-high school social contexts are hypothesized to be important in a sociological view of adult development because each one represents a major pathway to adulthood. The three pathways are conceptualized as social contexts not only because each one fulfills a crucial societal function[2] which brings the individual in contact with different

107

organizational goals, priorities, and social relationships, but because each context places quite different socialization pressures on the young person who enters it. If socialization is the "development or change that a person undergoes as a result of social interaction and the learning of new roles" (Gecas, 1979, p. 365), then it follows that the newcomer will likely be changed as he or she confronts the realities of context membership, learns new institutional roles, and becomes identified as a worker, a (college) student, or a serviceman, respectively.

## Occupational Choice, Achievement, and Ambition

The central theme that arises from the occupational choice literature is that it is a multidimensional phenomenon that is part of a larger process. The larger process in psychological theories centers on the development of vocational maturity (Super, 1957, 1970, 1984; Super, Starishevsky, & Jordaan, 1963) and occupational personalities (Holland, 1985). Stage theories typically underlie these approaches as rational actors move through decision-making stages or developmental stages. Under this paradigm, people maximize their occupational rewards by achieving the closest possible fit between their personality and their work environment. When a disjuncture between personality and work exists, people are motivated to seek alternative employment. The larger process addressed in the social psychological schema put forth by Blau, Gustad, Jessor, Parnes, and Wilcock (1956) is the emergent social and *structural* properties of society which have a powerful influence on the allocation of jobs and thus occupational "choice," such as broader educational and labor market opportunities. In this view, occupational choices are constrained within a larger system of relations (e.g., the social stratification system) that is outside the individual's control and of which he or she may be only partially aware (Rosenbaum, 1976). The social psychological approach to occupational choice includes a key structural component not found in the works of Super and Holland. In the former, opportunity structures such as industrial growth rates, economic conditions, employment policies, the historical milieu, demographic conditions, and high school tracking are incorporated along with such psychological attributes as personal preferences, attitudes, aptitudes, and personality types. Finally, the larger process described in Granovetter's (1973, 1974) structural model of occupational choice is one's location in an information network which can relay nontrivial and nonredundant job tips and recommendations.

The works of Super and Holland are reductionist in that both account for occupational choice exclusively in terms of psychological phenomena. Relatedly, Holland and Super fall victim to the ontogenetic fallacy of adult development (see Dannefer, 1984). On one hand, Super tries to explain occupational choice by reference to age-related developmental tasks associated with five life stages. Holland, on the other hand, largely discounts the impact of social context and social structure in occupational choice, except insofar as individuals seek occupations and occupational environments that are compatible with their personalities. This, of course, implies considerable knowledge of one's prospective work environment, which, it seems, is quite unlikely for most persons in the early adult transition period. Blau's social psychological schema of occupational choice avoids psychological reductionism and the ontogenetic fallacy by focusing attention on occupational choice as a function of individual characteristics being molded, shaped and "marketed" within a larger social structure and historical milieu. Thus, one's personality is important but it is only one of many factors influencing occupational choice and recruitment. Moreover, Blau et al. (1956) show that the transition to work may be controlled more by external forces than by personal attributes. At the very least, structural and personal factors work together in occupational decision-making. Blau's schema (he refuses to call it a theory) is especially useful because it places occupational choice in a multidimensional framework that is sensitive to the contributions of both psychology and sociology.

With the pioneering work on occupational choice framed in their social and psychological traditions, we present a post-high school context choice schema which attempts to identify the factors which lead one to enter the work force, the military or college after high school.

## Method

### Analytic Strategy

The choice schema is tested with multinomial logistic regression (MLR), a generalized linear modeling technique that is appropriate with a multiple categorical dependent variable and continuous and categorical independent variables (see Weisberg, 1985, pp. 260, 267–271; Aldrich & Nelson, 1984, pp. 30–47). MLR involves specifying a nonlinear probability model that is applicable to a polychotomous dependent variable.[3] The nonlinear probability model for a dichotomous dependent variable cannot be used in the polychotomous dependent variable case because the probabilities associated with the polychotomous contrasts may not sum to one (Aldrich & Nelson, 1984, p. 38). But, the dichotomous model can be extended, with some modifications, to the polychotomous case. Although other functions could presumably be used, the logistic function is typically used because it is far less computationally complex than the probit function—especially in the polychotomous case—and the results of probit and logit analyses are very similar (Aldrich & Nelson, 1984). The MLR routine in the statistical program LIMDEP 8.0 (Greene, 2002) was used to run the choice schema and to obtain maximum likelihood coefficients.

## Measures

There are four dependent variables and 25 independent variables. The four dependent variables are defined as follows: (1) The full-time work group ($n = 201$) consists of all boys who never served in the military, reported no more than one year of post-high school training or education, and were employed full-time one year after high school (Wave 4) and five years after high school (Wave 5). (2) The military group ($n = 263$) served in the Federal military for at least 18 months (the military's minimum for full service and benefits) between 1969 and 1974. (3) The college group ($n = 479$) never served in the military and reported at least 4 years of post-high school education by Wave 5. (4) An "other" group ($n = 607$) was constructed among all remaining subjects who could not be classified in a main social context. The "other" group is included to eliminate the possibility of sample selection bias (see Berk, 1983; Maddala, 1983). The independent variables consist of both social and psychological measures and are described in Table 3.1. All of the independent variables were measured during the first three waves of data collection (grades 10 to 12).

## Post-High School Social Context Choice Schema

### Specification of Choice Domains: Social and Psychological Factors

A range of individual-level and sociocultural factors that are hypothesized to bear directly upon the selection of work, military or college as a primary post-high school social context are now specified. The context

**Table 3.1. Factors Predicting Entrance into Work, Military, or College Contexts**

| Context choice predictors[a] | Response categories |
|---|---|
| *Family contingencies* | |
| First-born child | (1 = oldest, 0 = otherwise) |
| Number of siblings | |
| Farm family origin | (1 = from farm, 0 = otherwise) |
| Family socioeconomic status | |
| *Family and peer influences* | |
| Folks' want work after high school—10th grade | (1 = yes, 0 = no) |
| Folks' want college after high school—10th grade | (1 = yes, 0 = no) |
| Folks' happiness about entering military—12th grade | (−1 = unhappy, 0 = neutral, +1 = happy) |
| Peers' happiness about entering military—12th grade | (−1 = unhappy, 0 = neutral, +1 = happy) |
| Siblings' happiness about entering military—12th grade | (−1 = unhappy, 0 = neutral, +1 = happy) |
| Peers' respect for going to college—12th grade | (1 = not at all, 2 = somewhat, 3 = very true) |
| *School performance and experiences* | |
| Grade point average (GPA) in 9th grade | |
| College-bound high school track | (1 = college-bound, 0 = otherwise) |
| Vocational-bound high school track | (1 = voc-bound, 0 = otherwise) |
| Grade failure before 9th grade | (1 = failed one or more grades, 0 = otherwise) |
| *Attitudes toward self and society* | |
| Self-confidence—12th grade | (1 = low self-confidence, . . . , 5 = high self-confidence) |
| Dovish war attitudes scale—12th grade | (1 = low dove, . . . , 5 = high dove) |
| Hawkish war attitudes scale—12th grade | (1 = low hawk, . . . , 5 = high hawk) |
| Vietnam War dissent scale—12th grade | (1 = low dissent, . . . , 5 = high dissent) |
| *Ambitions and planfulness* | |
| Strength of college attending plans—12th grade | (1 = definitely won't attend, . . . , 4 = def. will attend) |
| Strength of votech attending plans—12th grade | (1 = definitely won't attend, . . . , 4 = def. will attend) |
| Have plans for the fall of 1969—12th grade | (1 = yes, 0 = no) |
| Ambitious work values scale—12th grade | (1 = low ambition, . . . , 5 = high ambition) |
| *Miscellaneous* | |
| Age | |
| Intellectual ability | |
| Worked during 12th grade | (1 = no, 2 < = 10 hours/week, 3 = > 10 hours/week) |

choice analyses are grouped under a conceptual framework involving six underlying choice domains: (1) family contingencies, (2) family and peer influences, (3) school performance and experiences, (4) attitudes toward self and society, (5) ambitions and planfulness, and (6) miscellaneous. The six choice domains and the context predictor variables subsumed under them are presented in Table 3.1.

## The Choice Domains

### Family Contingencies Domain

This domain consists of four objective variables hypothesized to influence context choice. This domain is important because of the significant role that families play in the formation of values and in status attainment (Gecas, 1981). The variables are most closely associated with the social structural variables described in the Blau et al. (1956) schema.

### Family and Peer Influences Domain

This domain is composed of six social psychological variables which gauge each boy's perception of family and peer expectations or perceived support for entering the three social contexts. These variables serve as important interpersonal influences which give shape and direction to a young man's post-high school choices.

### School Performance and Experiences Domain

This domain is composed of four objective school experiences which may impact a boy's context choices. These variables are important because they may constrain a boy's options (GPA, grade failure) or channel him into different career or educational paths (college- or vocational-bound tracks).

### Attitudes toward Self and Society Domain

This domain helps locate each boy's attitudes within a developmental or historical context. The domain involves two distinct psychological orientations. First, there is a measure of each boy's 12th-grade self-worth. If Super (1957) is correct in asserting that people attempt to choose jobs that implement their self-concepts, then self-esteem may well play a role in context choice. Second, as Blau et al. (1956) make clear, the sociohistorical milieu should be considered when assessing occupational choice, and by extension, context choice. Since the Vietnam War was on when most of the YIT boys left high school in 1969, contemporaneous attitudes toward that war and war in general are included as potentially important context predictors.

*Ambitions and Planfulness Domain*

This domain includes an assessment of the strength or certainty of each boy's educational plans, or whether he had any plans at all. This domain is informed by Super's (1957) concept of vocational maturity, which may have a direct impact on context choice.

*Miscellaneous Domain*

The miscellaneous domain is composed of three variables. Age is included because it has direct implications for a boy's draft risk. Intellectual ability is a very important individual attribute which may govern a boy's range of context options, particularly his suitability for college and also the military. High school work experience is included because it may indicate the degree to which a boy is oriented to work rather than further education (Finch & Mortimer, 1985).

A multidimensional schema of post-high school social context choice is presented which centers on 25 psychological and sociological variables, representing six broad content domains. The predictions are summarized in Table 3.2, below. A plus sign indicates that a factor is hypothesized to encourage entrance into a particular context, while a negative sign indicates that a factor is hypothesized to militate against entrance into a particular context. For example, the plus sign under the work column for the "farm family origin" variable in domain I indicates that a boy will be more likely to enter the work context if he was reared on a farm, while the negative sign under the military column indicates that being from a farm discourages military service. The absence of a sign indicates that no prediction was made.

## Factors Predicting Specific Context Choice

The factors that are hypothesized to lead to the choice of each of the three social contexts are now reviewed. While much has been written about the determinants of college entry, very little is known about the factors leading youths to enter the full-time labor force. Published information about military entrance is also limited.

*The Full-Time Work Context*

With respect to the family contingencies domain, there is evidence that family socioeconomic level (SEL), being raised on a farm, and having

Table 3.2. Summary of Social Context Choice Predictions

| Contexts context choice predictors | Post-high school social | | |
|---|---|---|---|
|  | Work | Military | College |
| *Family contingencies* | | | |
| First-born child | − | | + |
| Number of siblings | + | | − |
| Farm family origin | + | − | |
| Family socioeconomic status | − | − | + |
| *Family and peer influences* | | | |
| Folks' want work after high school—10th grade | + | | |
| Folks' want college after high school—10th grade | | | + |
| Folks' happiness about entering military—12th grade | | + | |
| Peers' happiness about entering military—12th grade | | + | |
| Siblings' happiness about entering military—12th grade | | + | |
| Peers' respect for going to college—12th grade | | | + |
| *School performance and experiences* | | | |
| Grade point average in 9th grade | | − | + |
| College-bound high school track | | | + |
| Vocational-bound high school track | + | + | |
| Grade failure before 9th grade | | + | |
| *Attitudes toward self and society* | | | |
| Self-worth—12th grade | | − | + |
| Dovish war attitudes scale—12th grade | | − | |
| Hawkish war attitudes scale—12th grade | | + | |
| Vietnam War dissent scale—12th grade | | − | + |
| *Ambitions and planfulness* | | | |
| Strength of college attending plans—12th grade | | | + |
| Strength of votech attending plans—12th grade | + | | |
| Have plans for the fall of 1969—12th grade | − | − | + |
| Ambitious work values scale—12th grade | − | − | + |
| *Other* | | | |
| Age | | + | |
| Intellectual ability | − | | + |
| Worked during 12th grade | + | | − |

a large number of siblings all play a role in entering the full-time labor force after high school. According to Gecas (1979) and Dreeben (1968), boys from low SES backgrounds are more highly disposed to non-college opportunities by virtue of family and school socialization, which should discourage post-secondary education and training in favor of a quicker placement in the labor force. Coming from a farm has been found to depress achievement ambitions (Spenner & Featherman, 1978, p. 390), which may in turn lessen the probability of seeking higher education. Sewell and Hauser's Wisconsin (1975, p. 154) data support the link between farm background and lowered college attendance. Increased sibship has also been linked to lower parental encouragement which in turn may reduce a child's aspirations (Spenner & Featherman, 1978, p. 390), grades, and academic self-concept (Johnston & Bachman, 1972, p. 112). Parents of large families also tend to be less encouraging of ambitious educational and occupational goals (Spenner & Featherman, 1978, p. 390; Blake, 1981, 1989, pp. 243–247).

In terms of the family and peer influences domain, boys who believe early in high school that their parents want them to go directly into the work force after high school may be more likely to do so. Such parental values may downplay academic achievement while emphasizing skills and experiences that enhance an adolescent's employment prospects, such as high school work experience, the development of vocational skills, compliant behaviors, and so on (e.g., Kohn [1969] see also Holland's [1985] discussion of how parents foster the replication of their personality types in their children).

Within the school performance and experiences domain, enrollment in a vocation high school track would indicate an ongoing intention to seek full-time employment after high school. A vocational curriculum might also frustrate college plans because of the socialization and the pressures (from other students, teachers, counselors, and even parents) which coincide with tracking (see Bowles & Gintis, 1968; Dreeben, 1968).

Within the ambitions and planfulness domain, boys with low work ambitions and some plans to attend vocational-technical school may be more inclined to enter the work force after high school. Persons with low ambition should be less likely to take the steps necessary for college entrance or to subject themselves to the rigors of college (or possibly the military), and might therefore be encouraged to make a swift transition to the workplace (e.g., Sheppard, 1973). However, since many of the skilled craft trades require or encourage completion of vocational training programs, those intending to enter the labor force after high school may opt for such training.

Finally, within the miscellaneous domain, intellectual ability and 12th-grade employment are hypothesized to influence entrance into the work context. Boys in the lowest mental ability groups should be more likely to be rejected for military service or college, thus making the labor force their most viable alternative. The 12th-grade employment variable measures the hours the boys worked in the spring of 1969. The literature suggests that high school employment tends to depress educational achievement, academic performance, commitment to school, and educational and occupational aspirations (e.g., see Mortimer, 2003; Mortimer & Finch, 1986; Finch & Mortimer, 1985). As such, employment during high school should be positively related to entering the workplace.

*The Military Context*

Within the family contingencies domain, family SEL and a farm family origin are hypothesized to be important military context predictors. Boulanger (1981, p. 501) reports that men from the two highest family SEL quartiles were far less likely to serve in the military during the Vietnam-era than were men from the two lowest quartiles. Johnston and Bachman (1972, p. 107) also found that attempted enlistment one year after high school rose as socioeconomic index sextile declined (*eta* = 0.125). Janowitz and Little (1965, p. 52) have noted that the military has long been recognized as a way for lower class men to advance themselves. Fligstein's (1980) study of military service entrance between 1940 and 1973 found that men from farm backgrounds were significantly less likely to serve throughout that period, with Vietnam-era farm boys being about 12.8% less likely to serve than were their nonfarm counterparts (p. 306).[4]

Within the family and peer influences domain, a boy's perception of parental, peer, and sibling attitudes toward him going into the military are hypothesized to influence his decision to enter or not. If he thought that these significant others would be happy if he entered the military, then he should be more disposed toward doing so.

Within the school performance and experiences domain, GPA, grade failure, and vocational education track are hypothesized to influence the decision to enter the military after high school. Boys with lower GPAs would be less able to go to college (Johnston & Bachman, 1972, pp. 87–88), which means that some other post-high school context would likely be chosen—such as the military. Rothbart, Sloan, and Joyce (1981, p. 136) found that Vietnam-era veterans had a history of lower school achievement than did their nonveteran peers. This history might lead a young man to choose the military as a time-honored way to win some legitimate respect and social approval in a rigorous nonacademic arena. Relatedly, the YIT boys who failed a grade were nearly three times more likely to try to enlist within a year after high school than were other boys (Johnston & Bachman, 1972, p. 91), which again speaks to the respect motive. Demographically, however, boys who failed a grade were older when they left high school and were therefore a higher drafted risk. Boys contemplating military service may also be attracted to high school vocational training in order to eventually use the service as a source of further technical training, and to improve their chances of landing non-combat jobs (i.e., truck mechanic versus infantryman).

Concerning the attitudes toward self and society domain, self-worth, opinions about the necessity of the Vietnam War, and attitudes concerning war in general are all hypothesized to influence entrance into the military. The self-concept motive of "compensation" (Rosenberg, 1978, p. 56) might play a hand in joining the military since the military expressly bills itself as a builder of confidence and pride. Concerning war attitudes, an exploratory factor analysis of 13 general war and specific Vietnam War attitudes revealed three factors: (1) Vietnam War dissent, (2) general hawkish war attitudes, and (3) general dovish war attitudes. A Vietnam War dissenter might be less inclined to serve in the military, while those expressing hawhishness might be strongly inclined to serve. Those who expressed general dovish war attitudes are also expected to have a higher probability of avoiding military service.

The ambitions and planfulness domain variables measuring whether a boy had *any* post-high school plans and his job ambitions may be important in choosing the military. Slightly more than 11% of the boys surveyed in 12th grade reported that they did not have any specific plans after high school and were "just going to wait and see what happens." Boys with no plans might enlist in higher numbers in order to avoid the uncertainties of civilian life at a critical point in their lives, which recalls Erikson's (1963) psychosocial moratorium, while those with plans would be expected to avoid the disruption of civilian plans imposed by military service (Suchman, Williams, & Goldsen, 1953). According to Janowitz and Little (1965, p. 54) boys "in the throes of intergenerational conflict and students without clear-cut goals are advised to join the service and 'grow up.'" A work ambitions construct is included because it reflects Janowitz and Little's (1965, pp. 52–53) "military laziness thesis" (see also Johnston & Bachman, 1972). According to the thesis, the military may be chosen by enlisted men who escape into the military in order to place "individual security ahead of competitive achievement" (p. 53).

Finally, within the miscellaneous domain, age is hypothesized to be an important military context predictor because older youths would be more susceptible to being drafted or to enlisting under threat of the draft, especially if they were not sheltered by a college deferment.[5] Approximately 17% of the YIT boys were drafted, while another 29% said they would have been drafted if they had not enlisted.

*The College Context*

Three factors within the family contingencies domain are hypothesized to influence the selection of college as the main post-high school context—first born child, number of siblings, and family SEL. The literature suggests that first born children have somewhat higher achievement ambitions than children with lower birth orders (Spenner & Featherman, 1978, p. 390; Blake, 1981). And higher achievement ambitions have been linked to increased attainment, which might initially be expressed as a higher probability of going to college (Spenner & Featherman, 1978). A supporting argument is that first born children might receive more sustained family support for entering and remaining in college because of so-called "first-mover" advantages (see Williamson, 1975, pp. 34–35 for an explanation of first-mover advantages). Family size might also influence context choice. If it is true that increased sibship depresses aspirations and lowers educational attainment (see discussion in work section, above), then the reverse should also be true: decreased sibship should raise a child's aspirations and therefore increase the probability of obtaining higher education (Blake, 1989). Family SEL is solidly linked to both achievement ambitions (see, e.g., Spenner & Featherman, 1978) and occupational attainment (e.g., Sewell & Hauser, 1975, pp. 105–106). It is also well known that children from higher SES backgrounds are far more likely to go to college than are other children (Borman & Hopkins, 1987).

Two variables in the family and peer influences domain are hypothesized to have a direct influence on going to college—whether a boy had an early perception that his parents wanted him to attend college after high school and how much a boy thought his peers respected a person for going to college (measured in 12th-grade). The influences of significant others, and especially parents, have been shown to be among the "most potent determinants of ambitions," particularly among adolescents (Spenner & Featherman, 1978, pp. 391–392; Blake, 1981, pp. 439–440). Furthermore, parental encouragement and expectations have strong direct and indirect effects on children's educational achievement (e.g., Kerckhoff, 1974, pp. 68–70; Sewell & Hauser, 1975, pp. 103ff). Parents who convey an early expectation that their son should attend college may not only give him a clear goal to pursue, but may also reward those behaviors and attitudes that will optimize the probability that he attend college.[6]

Within the school performance and experiences domain, high school GPA and enrollment in a college-bound track are expected to be closely associated with going to college. GPA is important on one hand because it indicates academic achievement, while on the otherhand boys with higher grades should be more likely to have higher educational and occupational aspirations, although Kerckhoff (1974, p. 38) and Sewell and Hauser (1975, pp. 92–96) report that the effect of GPA on aspirations is small when intellectual ability and socioeconomic factors are controlled. Spenner and Featherman (1978, p. 395) note that several studies have shown that enrollment in a college preparatory program has a modest net effect on educational and occupational ambitions. Indeed, Borus and Carpenter (1984, p. 99) found in their analysis of National Longitudinal Survey data from 1979 and 1980 that students who were enrolled in college preparatory tracks were about 28% more likely to enroll in college upon leaving high school.

Within the attitudes toward self and society domain, self-worth and attitude toward the necessity of the Vietnam War are hypothesized to influence choosing college over another context, especially the military. Self-worth is a theoretically interesting predictor because of its long link to aspirations and achievement (Gordon, 1972, pp. 77–79). A major finding of Gordon's monograph on the social psychology of adolescent achievement was that among middle, working, and lower class whites and among middle class blacks, the percentage of students aspiring to attain college degrees increased precipitously as self-esteem increased. Concerning war attitudes, Bachman et al. (1978, p. 136) report that as level of post-high school educational attainment increased, criticism of the Vietnam War also increased.

Thinking about the ambitions and planfulness domain, those headed for college are hypothesized to have specific short-range plans, more definite intentions to attend college, and ambitious work values. Since going to college necessitates specific planning, those planning to attend college should have many of their plans settled in the spring of their senior year in high school. The definiteness of a boy's college plans is also expected to influence the likelihood that he will attend. Among high school students surveyed in the 1979 and 1980 National Longitudinal Surveys, those who expected to attend college were significantly more likely to actually enter college shortly after leaving high school (Borus & Carpenter, 1984, p. 98). In terms of the link between ambitious work values and context choice, if it is true that many high status jobs are only attainable by at least going to college, then college-bound boys should be more ambitious than those not planning to attend college.

Finally, among the "miscellaneous" variables, intellectual ability and high school employment are hypothesized to be important factors in entering the college context. Many researchers have either found or postulate a link between increased intellectual ability and increased aspirations (e.g., Spenner & Featherman, 1978, pp. 402–403; Gordon, 1972, pp. 97ff) and achievement (e.g., Kerckhoff, 1974, p. 44ff; Sewell & Hauser, 1975, p. 115; Lowman, Galinsky, & Gray-Little, 1980, pp. 74–75). Recent research indicates that students who go to college directly after high school are more likely to be from the two highest SES and intellectual ability quartiles (U.S. Department of Education Statistics, cited in Borman & Hopkins, 1987, p. 4). Employment during high school has also been shown to decrease college attendance (Finch & Mortimer, 1985).

## Findings

Attention is now turned to the significant MLR results obtained for the work, military, and college contexts only (see Table 3.3). A residual "other" group was included in all of the analyses in order to avoid selectivity bias, but the findings for this group are not reported here for the sake of simplicity. (See Appendix A for the complete MLR results.) Although the results are represented as choice-pairs, they should not be mistaken for dichotomous logit results. In the polychotomous dependent variable situation, a solution is achieved by selecting one context as the baseline group from which competing choices among the remaining alternatives are compared. Since I am interested in the full array of context choices, three sets of logistic regression analyses were performed, with the redundant contrasts excluded.[7] The context listed on the top of each column in Table 3.3 is the first object of comparison. Thus, the sign of the $b$ coefficient for the siblings variable is −0.34 for the college versus work comparison (column 2), which means that as the number of siblings increased, the *probability of entering the college context rather than the work context decreased*, after controlling all the other variables. Conversely, as the number of siblings increased, the probability of entering the work context as opposed to the college context *increased.*

The boys who chose work or the military as their main post-high school social contexts had some similar characteristics, and those who chose college or were in the "other" group were similar to each other. Overall, the work context was chosen by boys from large families who were in the lower socioeconomic strata, were enrolled in high school vocational tracks, and were working more hours in their senior year, as compared to those in the college group. The workers tended to have the lowest intellectual ability

**Table 3.3. Results of Multinomial Logistic Regression Analyses, Summarized**
**($N = 1,082$) (Unstandardized Maximum Likelihood Coefficients)**

| Independent variable | Military versus work b | College versus work b | College versus military b |
|---|---|---|---|
| *Family contingencies* | | | |
| First-born child | | | |
| Number of siblings | | −0.34*** | −0.41*** |
| Farm family origin | −0.91* | | 1.50*** |
| Family socioeconomic status | | 0.46*** | 0.31*** |
| *Family and peer influences* | | | |
| Folks want work after high school—10th-grade | −0.90* | | |
| Folks' happiness about entering military—12th grade | 0.51** | | −0.93*** |
| Peers' respect for going to college—12th grade | 0.57** | 0.74*** | |
| *School performance and experiences* | | | |
| Grade point average in 9th grade | | 0.08*** | 0.08*** |
| College-bound high school track | 0.93** | 1.05*** | |
| Grade failure before 9th grade | 1.81*** | | −1.70*** |
| *Attitudes toward self and society* | | | |
| Hawkish war attitudes scale—12th grade | 0.73** | | |
| *Ambitions and planfulness* | | | |
| Strength of college attending plans—12th grade | | 1.11*** | 1.22*** |
| Strength of votech attending plans—12th grade | | −0.46** | −0.43** |
| *Other* | | | |
| Intellectual ability | 0.09* | 0.15** | |
| Worked During 12th grade | | −0.29* | −0.29* |

*Note*: The sign indicates whether the first context listed was more likely to be entered given the particular predictor variable (indicated by a positive $b$), or whether the second context was more likely to be entered (indicated by a negative $b$). Model Chi-Square = 784; df = 75.
*$p < .05$　**$p < .01$　***$p < .001$.

and to have friends who were the least impressed by going to college than members of any other context. Boys who early on believed that their parents wanted them to enter the work force after high school were significantly more likely to enter the work context over the military context. These outcomes were all predicted (see Table 3.2). Two other predictions were not confirmed. The work force-bound boys were no different in ambitious work values or general planfulness than boys headed for the other contexts.

The military context, like the work context, was chosen by boys who tended to express little desire to attend college, were generally from large families in the lower socioeconomic strata, were enrolled in a vocational track, and were poorer students when contrasted to the boys headed for college. The military-bound, as opposed to those headed for one of the other contexts, were most likely to come from nonfarm backgrounds, to have failed a grade, and to believe their parents would be happy if they served in the military. The military group tended to express more hawkish attitudes than the work group, but was not significantly more hawkish than the college-bound group. Eight predictions were not confirmed. These data indicate that the opinions of the military-bound boy's peers concerning military service had no net effect after controlling other, especially parental, influences. Level of self-worth was not a significant factor in seeking the military as predicted, thus failing to confirm Rosenberg's "compensation" motive in this instance. Also, attitude toward the Vietnam War did not seem to be a significant factor in entering the military context. The military laziness thesis and "psychosocial moratorium" hypothesis were not found to significantly influence entrance into the military. Finally, age was not a significant factor in context choice.

The college context was chosen by boys who came from the smallest families and the highest socioeconomic status backgrounds, had the highest GPAs and were in a college track, had a strong intention of going to college, and worked the least amount of hours in 12th-grade. Boys with higher intellectual ability and friends who were impressed by going to college were more likely to go to college as opposed to work. Contrary to prediction, being college-bound was not significantly predicted from being a first born child, expressing awareness of early parental desire to attend college, possessing high self-worth, expressing higher general planfulness, or being occupationally more ambitious than boys in other contexts.

## The Class of 1969 in Sociohistorical Perspective

> In the 1950s and 1960s, our country had to find a way to house and educate the 'baby boom' generation. In the 1970s and 1980s we had to integrate this generation into the labor force. In the next century, we will have to support them in their old age and provide for their health care.
>
> NANCY-ANN MIN DePARLE, *Administrator,*
> *Health Care Financing Administration, U.S.*
> *Department of Health and Human Services*
> *in testimony before the Senate Special*
> *Committee on Aging, February 18, 1998.*

### *Introduction*

The purpose of this section is to place the high school class of 1969 (or birth cohort of 1951) in sociohistorical perspective. Rather than paint this portrait from the various demographic data contained in the YIT dataset, I want to draw on census data, appropriate popular media of the day, some personal recollections, and a smattering of other sources. The portrait is not intended to be exhaustive, nor do I intend it to be or pretend that it is definitive. Instead, it is designed to put the young men's cohort into the broader context of American society at various points in their lives, from the 1950s to the mid-1970s, when the YIT study ended.

Naturally, this is not a one-size-fits-all portrait. The boys from the class of 1969, and those sampled in the YIT study, are as diverse as the society from which they arose. And it must be remembered that because of sampling and the sociopolitical conditions of the day, my focus in this book is necessarily on white boys (see chapter 2 for a full explanation). Consequently, we have boys from the South and boys from the North, West Coasters raised on the car culture of the 1950s and 1960s, and Northeast city kids who never felt any particular need to get a drivers' license because mass transportation was the only viable means of getting around. Some of the boys were born to highly compensated professionals and executives while others sometimes went to bed hungry as children or their families lived hand-to-mouth. Some blue-collar children never ate in a restaurant because their parents deemed it either an unaffordable luxury or a frivolous waste or hard earned money.

Many grew to young adulthood without ever knowing a person from another race. Some of the kids from the class of 1969 were raised in inner-city ethnic neighborhoods that started disappearing in their teen and early adult years. The causes of these declining ethnic neighborhoods are many and familiar. Some of their parents moved to the suburbs to avoid bussing or living in proximity to African Americans after the Civil Rights Acts of 1964 and 1965 outlawed real estate discrimination. Some left after all their children were grown. For others, the family home became too much of a burden for their aging parents, and still others left because they started feeling unsafe in the neighborhood where they raised their children and may have been raised themselves.

## *Family Life*

In 1950, American women averaged 23.6 live births per 1,000 in the population (U.S. Bureau of the Census, 1965). Ten years earlier, in 1940, the rate was 17.9, and 10 years later, in 1960, it rose slightly to 23.7. After the early 1960s it would continue to drop for white women. When the boys in my study were born, the marriage rate was relatively high (11.1 per 1,000 in the population) and the divorce rate was comparatively low at 2.6/1,000). While the divorce rate would hold steady well into the 1960s, the marriage rate would start a slow decline, until it evened out in the mid-1960s. Slightly more than two-thirds of American adults were married in the early 1950s (U.S. Bureau of the Census, 1956), with the proportion rising slightly as the military shrank after the Korean War. (Military service is a disincentive to marriage, especially among those not planning a career in the service.)

The boys from the class of 1969 were born to relatively young parents, a trend that began in the late 1800s but accelerated sharply after World War II. Over half of their fathers married before age 23 while over half of their mothers married before age 20 (U.S. Bureau of the Census, 1956). While the average household size in 1950 was 3.54 people, a slight decline from a decade earlier (3.76 in 1940), the average number of children under 18 per family increased from 1.25 in 1940 to 1.32 in 1956, the year the 1969ers were getting ready to start kindergarten (U.S. Bureau of the Census, 1956). These trends reflect the high birth rate following World War II and the decline of married couples living with others (i.e., "doubling"). In the pre-war years, the nation's housing supply was much more limited and the economy certainly weaker than the post-war years. This forced many couples—especially newlyweds—to double in another's household, often a parent's. Indeed, the proportion of married couples without their own home or apartment decreased sharply in the 1950s and continued that trend into the mid-1960s, when only 2 percent of married couples lived with others. Also, between 1950 and 1965, the proportion of families with no children declined substantially while the proportion with three or more children rose from 1 in 6 in 1950 to 1 in 4 in 1965 (U.S. Bureau of the Census, 1966). During the class of 1969's childhood and adolescence, most households consisted of a husband and wife, with the husband the primary breadwinner (U.S. Bureau of the Census, 1966).

Also during this time, the migration from farm-to-city continued. When the boys were born, most of the country's population growth was registered in metropolitan areas; by the time they were in their early teens it had shifted to the country's mushrooming suburbs (U.S. Bureau of the Census, 1966).

## *Schooldays*

Nearly all of the boys from the class of 1969 were born in 1951 and started kindergarten in the fall of 1956. The general policy of the day was to start kindergarten (if your state required it) the year you turned five or to start $1^{st}$ grade the year you turned six. Approximately 58.9 percent of five year-olds in the fall of 1956 were enrolled in a public or private school, by age six, that figure rose to 97 percent (U.S. Bureau of the Census, 1957). In the mid-1950s, most schools started right after Labor Day, the unofficial end of the summer. Many city-born kids walked to school or took public transportation. The Minneapolis Public School System, for example, did not own any school busses until the 1970s, when school desegregation was implemented and children needed transportation to schools far away from their neighborhoods.

If the Baby Boom Generation can be roughly defined as all American children born between 1946 and 1964, then the boys in my study are in the first third of that generation. In 1956, approximately 3.6 million children were enrolled in $1^{st}$ grade (Dept. of Health, Education, and Welfare, 1964). In the fall of 1963, when the boys were halfway through their required schooling ($7^{th}$ grade), 46.9 million pupils were enrolled in America's public schools (Dept. of Health, Education, and Welfare, 1964). When they ended kindergarten in 1957, 14.1 percent of all children were enrolled in private schools (U.S. Bureau of the Census, 1957). Reflecting birth rates shortly before and during World War II, approximately 2.8 million Americans typically reached age 18 annually (the age when most people start college) and jumped to 3.7 million by 1965 (U.S. Bureau of the Census, 1966), just as the boys were approaching college-age themselves. Coextensive with this trend was a burgeoning of the number of young people entering college itself. This was brought on, in part, by post-war prosperity, the initiation of government guaranteed student loans, the massive expansion of community and state colleges across the nation, and the lure of college deferments to the Vietnam-era draft.

## *The Economy*

The economy was generally good when the boys were born and remained strong for the next two decades. "Made in America" was a frequent sight on all manner of consumer products and was both a source of pride and a symbol of excellence. The seasonally adjusted unemployment rate in 1951 averaged a healthy 3.2 percent, a substantial decline from the immediate post-war years, and a considerable drop from just a year earlier when it averaged 5.2 percent (U.S. Department of Labor, 2004). This propitious decline in unemployment was due in large part to the economic effects of the Korean War. Over the next 20 years, the rate fluctuated considerably until the mid-1970s. From about 1975 until the early 1980s, unemployment hit post-war records of double-digits. This is important, because just as the boys were reaching adulthood and seeking a permanent place in the labor force, the economy took a downturn that would last a decade.

During much of the boys' childhoods, inflation was relatively low, but it jumped dramatically in their early adulthoods. In 1951, the inflation rate was 7.64. When they entered junior high in 1964, it was a low 1.37. By the time they left high school in 1969, it had increased to 5.29. However, at the time of the last YIT interview, in 1974, the year most turned 23, it spiked to a whopping 11.11 (Economic History Services, 2004).

Consumer patterns during their childhood showed both stability and change. In 1950, Americans spent 11.2 billion dollars on all forms of recreation. By 1960, the year most of them turned 9 years old, 88 percent of American households owned a television set. That figure would rise slightly to 94 percent before they left high school (U.S. Department of Commerce, 1967). In contrast, as the consumer economy hummed along in the 1960s, the rock-and-roll era was in full swing, and 45 RPM and LP records became mainstays of home entertainment. The proportion of households with radio and phonographic equipment went from 22 percent in 1960 to 35 percent by the end of the decade. Although the majority of households had at least one car throughout the 1960s (75% on average), by the time the boys left high school the number of households with two or more cars jumped from 15 percent in 1960 to 25 percent at decade's end (U.S. Department of Commerce, 1967). Many of the cars were given to or purchased by the teenagers in the household.

Still, a majority of households went without what many today consider basic necessities. Dishwasher ownership was a mere 7 percent in 1960 and only 11 percent by decade's end. Air conditioning was still out of most Americans' reach in the 1950s and 1960s. In 1960, 14 percent of households had the luxury of cooled air, and only 20 percent had it when the boys left high school. Finally, the mothers of the boys still routinely dried their laundry on a clothesline or hauled them to a nearby Laundromat.

At the beginning of the decade, approximately 17 percent of households had a clothes dryer, but the number nearly doubled by decade's end, to 30 percent (U.S. Department of Commerce, 1967).

## *Popular Culture*

Although the boys from the class of 1969 missed the golden years of radio, they were on the ground floor of the new mass medium: television. As toddlers and pre-kindergartners, they got some anticipatory socialization from *Ding Dong School* and its host Miss Frances (actually Dr. Frances Horwich). Other favorites of young children were *Captain Kangaroo* and the cartoons, "Tom Terrific" and "Mighty Mouse" on Saturday mornings.

Adventure shows were everywhere on television, from Westerns like *The Lone Ranger, The Rifleman, and Wagon Train*, to contemporary fare such as *Sky King, Sea Hunt*, and *My Friend Flicka*. My two favorite variety shows in the 1950s and early 1960s were *The Ed Sullivan Show* and *The Red Skelton Show*. The day after every evening telecast of *Red Skelton*, my friends and I at Lexington Elementary School would relive nearly every minute of the show as we made our way through the lunch line and then to our table. The next 40 minutes were spent in heaves of laughter as we tried to top each other with quotes and gags from the night before.

Many of the television shows in the 1950s and 1960s carried pronounced social and moral messages. We were taught to stand-up to bullies, try to get along with everyone, and if necessary sacrifice ourselves for others, especially if honor was at stake. *Branded* typified this ethic. Week after week we watched the star get literally drummed out of the army. Although it was a ruse that enabled him to infiltrate villainous enclaves and bring them to justice, the contempt of the decent people he was continually forced to endure was all too real. *Combat* was a huge hit among pre- and post-pubescent boys of my generation, as were *The Rifleman, Have Gun Will Travel, Davy Crockett, Gunsmoke, Maverick, Rin Tin Tin,* and *Wanted: Dead or Alive.* If these morality tales were not sufficient, there were plenty of crime shows on television to make us think twice about breaking the law, especially the likes of *Superman, Dragnet, Highway Patrol,* and the *Naked City.* And patriotism was displayed nearly everywhere, if not in the aforementioned then certainly in *West Point Story* and *Men of Annapolis.* Glimpses of near domestic perfection could be seen in *Father Knows*

*Best, Leave It To Beaver, The Danny Thomas Show,* and *The Donna Reed Show.*

Children of the 1950s and 1960s generally saw movies in three venues: television, drive-ins, and Saturday matinees. For as little 25 cents, a kid could be entertained all afternoon, and for a dime more, get popcorn. On Saturday, November 2, 1957, when the class of 1969 was in 1st grade, *The Indianapolis Star* lists the following movies appropriate to young children showing at area movie houses: *The Ten Commandments,* Red Skelton in *Public Pigeon No. 1, Frances in the Navy,* Walt Disney's *Man in Space,* and *Around the World in 80 Days.* Several theaters list their Saturday matinees as simply "Matinee," "Matinees Today," or "Cartoons." The specifics seemed irrelevant; the kids knew they were in for a treat no matter what, and the parents knew the kids would be out of their hair for at least two hours. On Saturday, November 1, 1969, the fall after graduation, the 1969ers could pick from a variety of anti-hero movies or more risqué fare. Again, *The Indianapolis Star* lists the following movies that an 18 year-old male might like to see at area theaters with his buddies or a date: *Easy Rider; Medium Cool; Those Daring Young Men in Their Jaunty Jalopies; Bonnie and Clyde*; the testosterone fueled*Bullitt, Private Navy of Sgt. O'Ferrell* with Bob Hope and Phyllis Diller; *Barbarella; Fraulein Doktor*, about a wily female German spy in World War I, or *The Great Train Robbery.* In 1974, when the YIT study ended, Indianapolis theaters were showing *Airport 1975; Return of the Dragon; Soldier Blue* about a U.S. cavalry massacre of Cheyenne and a young soldier's turn of heart; *Savage Sisters*, a "blaxploitation" film about urban guerrillas; *The Odessa File*; and *Foxy Brown*, with the ad warning "Don't mess around with Foxy Brown she's the meanest chick in town!"

What is the point of all this? Quite simply, the media gives us a glimpse of societal values and interests—or hopes—at various points in time. From the patriotic innocence of the 1950s to the seriousness of the late 1960s and early 1970s, when much of American was shaking and cynical, or just bewildered, movies and television give us another view of what people might be thinking about and feeling.

### *Summary*

The boys from the class of 1969 witnessed some of the nation's greatest triumphs and deepest tragedies. Many felt supremely confident of America's moral and military might, until the early 1970s when national and international events severely challenged such pretensions. For the nation's Catholics, John Kennedy's presidential election was enormous. Aside from family photos, many Catholic households proudly displayed two pictures in places of honor, the pope's and Kennedy's. The 1969ers were in 7th grade when his assignation rocked the country. Shortly afterward the Beatles took America's teenagers by storm, and we were in complete thrall when a new space mission was launched. Many of the boys from the class of 1969 were only vaguely aware of the seriousness of the Vietnam War in 1965. It seemed like small potatoes compared to World War II. All that changed as the American death toll escalated and the war's coverage dominated the national news, especially dinnertime TV news.

Surrounding this was a time of tremendous political achievement in the United States. In the summer of 1965, the house and senate debates of the pending civil rights legislation was carried live on network television. That summer also saw the start of some of the worse race riots in American history. Later, the 1968 Tet Offensive in Vietnam would jolt the country and begin the slow, painful American withdrawn from that war. In 1969, just after graduation, Woodstock took place in upstate New York. Even if you had no clue it was going on at the time, when the documentary of the same name came out the next year, the 1969ers flocked to see the movie and buy the soundtrack.

If one of the boys went to college after high school, in May of his freshman year he was likely exposed to the tumult created by the U.S. invasion of Cambodia and the massive student unrest—and killings—that followed. Two years later, tear gas would once again pervade many college campuses as students who thought the Vietnam War was winding down became enraged when the U.S. mined North Vietnam's Hiphong Harbor. It was right about this time that many in the anti-war movement started referring derisively to blue-collar workers who they presumed were pro-war as "hard hats." The terms later got generalized by others to anyone not well educated and with a parochial, "conservative" political outlook. The boys from 1969 who were primarily full-time workers after high school would have been exposed to some of this misplaced contempt for manual laborers.

The boys from the class of 1969 thus experienced more social, political, and economic change and turbulence in their first two decades of life than most other generations. The consequence was a nation and cohort deeply divided along ideological and experiential lines. Those in the military experienced a formally proud army in disarray. The workers bene-fited from a strong economy and cheap energy (until 1973) that afforded a variety of semi-skilled and skilled industrial jobs that generally paid well. The college-bound found themselves in vibrant institutions full of ideas and criticisms of the status quo, but with amble opportunity to have a ball behind their safe and comfortable student statuses.

## Summary and Conclusions

This chapter began by developing a general schema for identifying the important factors influencing a boy's decision to enter the full-time labor force, the military, or college after high school and concluded with a sketch of the sociohistorical context of the class of 1969. Concerning the former, the literature review indicated that context choice, like occu-pational choice, has multiple determinants. Twenty-five potentially impor-tant context choice predictors were identified, ranging from psychological constructs such as intellectual ability and self-worth, to social locational variables such as family socioeconomic status and farm versus nonfarm origin. The 25 predictors were grouped into six influence domains dealing with family background characteristics, family and peer influences, school performance and experiences, attitudes toward self and society, ambitions, and other characteristics (see Table 3.1). Table 3.2 summarizes the context choice predictions made on the basis of the literature reviewed. In all, 14 variables were significant predictors ($p < .05$, see Table 3.3). It was also found that while every influence domain contributed to context choice, the family contingencies domain, the school performance and experiences do-main, and the family and peer influences domain tended to have the most influence, while the attitudes toward self and society domain had the least influence (see Table 3.3).

## Notes

1. Since choice of context is much broader than choice of occupation, and because college, and very often the military, are transitional involvements, it is helpful to view context choice partly as a function of achievement ambitions. Spenner and Featherman (1978) take a social psychological perspective on achievement ambition by placing it within the larger context of role theory and socialization theory. Briefly, role theory views the self as a complex set of roles that a person actually enacts or anticipates enacting (p. 384). Achievement ambitions are generally thought to arise through socialization to actual or

expected roles. Achievement ambitions are one aspect of a larger set of beliefs, skills, knowledge, and so on that are produced within the self (pp. 384–385). It is believed that these ambitions evolve throughout the life course in response to the socialization contexts and influences the individual encounters (pp. 384–385).

Achievement is broadly defined as "worldly success" (p. 384). It is measured through academic grades and credentials, income, occupational prestige, and so forth (p. 374). Achievement may be gauged in terms of two components: role residing or incumbency and level of role performance or accomplishment (p. 374). Role residing achievement is composed of the social evaluations, sanctions, and rewards a person receives through the roles that he or she assumes. Occupational roles, for example, entitle the incumbent to different "degrees of interpersonal deference (prestige), and more generally to [different] levels of renumeration, job security, and other rewards" (p. 374). Level of role performance indicates the degree to which a person meets or exceeds the expected standards of his or her role (p. 374).

Ambition is defined as a psychological orientation held with respect to achievement in and through roles (p. 374). As an attitude about self in "relation to specific sets of objects in achievement situations," ambition consists of these referents (pp. 374–375):

a. *Cognitive categories* the individual uses to perceive his or her role residing and role performance, (e.g., a job's status and financial rewards), and how intelligent and competent one's role performance is or is expected to be.

b. Affective*states* the individual feels because of role residing and performance. These states are manifested as pride, shame, fear, anxiety, and so on.

c. *Behavioral intentions* associated with one's ambitions, such as attending school, entering the labor force, raising children, and so on.

2. The basic societal functions of the work, military, and college social contexts are the production of goods and services, national defense, and the education and training of the nation's youth.

3. A linear probability model such as $Y_i = b_0 + b_1 X_i$ is inadequate because while the observed $Y_i$ is bounded between 0 and 1 (as when a context is chosen or not chosen), $b_0 + b_1 X_i$ is not so bounded (see Weisberg, 1985, p. 268; Aldrich & Nelson, 1984, pp. 12–14, 27–30). Furthermore, the assumption of nonlinearity is reasonable because Weisberg (p. 268) reports that S-shaped curves are a common outcome when modeling a binomial response as a function of predictors. According to Aldrich and Nelson (pp. 27–30), by incorrectly assuming linearity, one risks obtaining least squares estimates which, among other problems, will have unknown distributional properties resulting in invalid hypothesis tests and confidence intervals which may seriously misstate the magnitude of the true effects of X on Y.

4. An important reason why farmers tended to serve less is that farming was classified as an essential civilian occupation in selective service law, which could excuse some farmers from the draft. Farm families might have also discouraged their sons from enlisting in order to retain a valuable source of labor.

5. For instance, the class of 1969 is made up mostly of boys born in 1951 (who turned 18 in 1969). Given selective service law in effect at that time, the earliest a boy born in 1951 could be drafted would have been 1971, the year in which the entire birth cohort turned 20 years old. Starting in 1971, when the draft lottery was in effect, a boy without a draft deferment was at risk for conscription for only that year, after which he would no longer

draftable. (Boys who obtained college deferments would have largely avoided the draft altogether since the draft was abolished after 1972; incidentily, the draft ended with the 1952 birth cohort.) Boys born *before* 1951, however, would have been eligible for the draft as soon as they turned 19 and left high school. Men born before 1951 who were not draft deferred would have been draftable through 1970. The mean age for whites in the 1969 YIT sample was 18 years, with the maximum being 21 and the minimum being 17.

6. Although large amounts of missing data prevented the use of a parents' college expectation variable measured when the boys were 12th-graders (Wave 3, 1969), a parallel measure of perceived peer respect for going to college in grade 12 is included in the choice analyses.

7. First, the work group was used as the baseline for comparison: namely, contrasting work to the military, to college, and then to the "other" groups (see Aldrich & Nelson, 1984, p. 38). The choice between the work context versus the military context is a simple reflection of the choice between the military context versus the work context. The only difference between the two contrasts is in the *sign of the b coefficients* associated with the various predictors. Second, the military group was used as the baseline, although only the college versus military contrasts are reported because the work versus military contrast is redundant (see column 1). Finally, the college context was the baseline, but since the only new contrast was between "other" versus college, the results are not reported.

# 4

## Pathway's Effects on Self-Esteem

### Overview

This chapter examines the effect of three different adult social contexts—the workplace, the military, and college—on a person's self-esteem. The theoretical rationale for the investigation is provided by three perspectives: (1) Dannefer's (1984) sociogenic thesis, (2) Glenn's (1980) aging-stability hypothesis, and (3) Bronfenbrenner's (1979) ecological perspective on human development. Dannefer (1984) observes that many models of adult development contain an 'ontogenetic fallacy' that "surrender[s] ... socially produced age-related patterns to the domain of 'normal development'" (p. 101). Rather than use a sociogenic approach to examine the profound influence diverse social environments have on psychological development over the life course, some research still views development ontogenetically. With the approach: (1) the individual is treated as a self-contained entity, and both the (2) "profoundly interactive nature of self-society relations" and (3) the complexity and variability of social environments go largely unexplored (p. 100).

The aging-stability hypothesis (Glenn, 1980) complements the sociogenic thesis: "attitudes, values, and beliefs tend to stabilize and to become less likely to change as a person grows older" (p. 602). Although the particular timing of the decline in "change-proneness" over the life course remains unsettled, "it is generally accepted that late adolescence and early adulthood ... are times of high susceptibility to change" (Mortimer, Lorence, and Kumka, 1986, p. 7). However, Glenn (1980) and many other life course researchers reject the idea that development ceases with the attainment of adulthood. The adult development literature also suggests that adults are most subject to change when they enter new organizational contexts or move across organizational boundaries, or when salient social roles are acquired or relinquished (e.g., Bush and Simmons, 1981; Mortimer and Simmons, 1978; Van Maanen, 1976; Van Maanen and Schein, 1979; Bronfenbrenner, 1979). Glenn stresses the need to assess psychological stability and change through measures of attitudes with stable or highly abstract objects (1980, p. 605; see also Sears, 1981, p. 184). The self indeed is such an object (e.g., Rosenberg, 1979; Wylie, 1974; Wells and Marwell, 1976; O'Malley and Bachman, 1983), and is thus appropriate to the "sociogenic" and "aging stability" models.

Bronfenbrenner's (1979) ecological perspective places the developing individual within a nested system of micro to macro relations which impact the person directly and indirectly. The microsystem is the immediate setting—or "context" in the present research—containing the developing person. The mesosystem is the relations between microsystems, while the exosystem reflects the events of the larger social system (e.g., war, civil unrest, economic depression) which affect the developing individual.

I employ Murphy's (1947, p. 996) concise definition of self-concept: "the individual as known to the individual." Self-esteem is the evaluative component, one's assessment of the worth of the self as an object (Owens, 2003; Rosenberg, 1979; Wells and Marwell, 1976). However, as discussed in chapter 2, the dimensionality of self-esteem is unsettled. While Rosenberg finds it a unidimensional, or global construct, others (e.g., Kohn, 1969, 1977; Kohn and Schooler, 1983; Owens 2001, 2003, 1994, 1993) separate it into its positive and negative self-evaluative components. Kohn (1977) views self-confidence as the "positive component of self-esteem: the degree to which men are confident of their own capacities" (p. 81) and self-deprecation as the "self-critical half of self-esteem: the degree to which men disparage themselves" (p. 82). This empirical division of self-esteem enables one to be "simultaneously confident of one's capacities and critical of oneself" (p. 82).

Young persons in the early adult transition period are simultaneously forming their adult personalities and assuming new social status positions (Erikson, 1968; Levinson, 1978). They are making seminal ecological transitions—shifts in roles and settings (Bronfenbrenner, 1979). In American society, males, roughly between ages 17 and 22 (Levinson, 1978), have essentially three choices upon leaving high school (Panel on Youth, 1974, p. 67): they can enter the full-time labor force, enter the military (initial service typically ranges from two to four years), or continue on to post-secondary schooling. Some combine these, switching, for example, between college and the labor force. My focus here is on youths who made a primary commitment to either work, the military, or college. One's context status (e.g., worker, soldier, student), and socialization to its role demands, centrally influence self-concept development (e.g., Linton, 1936; Mortimer and Simmons, 1978; Bronfenbrenner, 1979).

The present research examines the broad effect of overall social context on self-esteem, which has generally escaped thorough analysis, rather than *intra*-context processes. Focus on *inter*-context processes admittedly paints the sociogenic thesis and the aging-stability hypothesis broadly, but it also affords a straightforward assessment of the role these "ecological transitions" play in young adult development.

## Social Context and Socialization

### The Vietnam Era Military

Military service entails the highly structured social controls of a total institution (Goffman, 1961; Janowitz, 1971; Moskos, 1970). Reminiscent of Foucault's (2003) notion of a panopticon, pervasive surveillance of individuals, the modern U.S. military ensnares its inhabitants in a vast network of impersonal power relations. The military is also a powerful agent of socialization and sometimes resocialization (e.g., Gecas, 1981; Eisenhart, 1975; Faris, 1976; Moskos, 1976) that may have both positive (e.g., Elder, 1974, 1986; Martindale and Poston, 1979) and negative (e.g., Santoli, 1981; Egendorf, Kadushin, Laufer, Rothbart, and Sloan, 1981) impacts on personality and status attainment, as manifested upon return to civilian life. Above all, the military demands compliance to its rules and customs on both a formal and informal level (e.g., Stouffer, Suchman, DeVinney, Star, and Williams, 1949; Bombard, 1981; Andreski, 1968; Moskos, 1970; Janowitz, 1971). This resocialization is easily seen in every recruit's extensive indoctrination regardless of civilian background, branch of service, or whether headed for Officer Candidate School or the motor pool. Topics covered range from military courtesy (formalized rules for rank-based interactions) to military law and custom. In short, recruits are intensively socialized to narrowly defined roles and statuses within a monolithic bureaucracy (Andreski, 1968; Segal and Segal, 1983).

Despite the population's large proportion of veterans, this powerful institution's enduring social and psychological effects have yet to undergo vigorous empirical examination. Elder (1986) summarizes this well in a passage decrying the paucity of research on the effects of World War II and the Korean War on the life course of veterans. The same argument holds for Vietnam.

> Military service represents a pervasive and varied experience ... across successive birth cohorts .... Most Americans have passed through a time of war en route to more schooling, a good job, marriage, and children. Some entered the Armed Forces by choice or conscription—over 20 million in World War II and the Korean War. Others knew only family members or friends who served, the survivors, and casualties. Still others made life-shaping decisions in terms of the military draft, from postponing marriage, schooling, and careers to becoming conscientious objectors. Considering all of these influences, it is striking how little is known about their consequences in lives and across the generations (Elder, 1986, p. 233).

Military service has impacts that roughly divide into two subareas—normal psychological development and severe psychological distress. War stress and trauma research consistently shows a direct relationship between heavy combat exposure and later symptoms of severe distress, such as Post-Traumatic Stress Disorder (e.g., Laufer, Gallops, and Frey-Wouters, 1982; Laufer, Yager, Frey-Wouters, and Donnellan, 1981; Kadushin, Boulanger, and Martin, 1981; and Hendin and Pollinger-Haas, 1984).

A literature review uncovered very little published scientific evidence about the impact of Vietnam-era military service on self-esteem. Two well-done longitudinal studies, however, suggest that, in general, military service had little affect on self-esteem (Kanouse, Haggstrom, Blaschke, Kahan, Lisowski, and Morrison, 1980; Bachman, O'Malley, and Johnston, 1978). However, neither study thoroughly controls for selection processes or time-in-service. In a report based on National Longitudinal Study of the High School Class of 1972 (NLS-72) data, Kanouse and associates (1980) also find no relationship between military service and changes in self-worth[1] four years after high school, with ability, family socioeconomic status, high school rank, and other controls (p. 39). Bachman and associates (1978, pp. 126–127) also find military service is not apparently related to later global self-esteem.[2]

The earlier findings imply that Vietnam-era military service had little impact on self-esteem, except for the sometimes deleterious effect of heavy combat, which neither Kanouse and associates (1980) nor Bachman and his team (1978) assess. However, many previous studies do not account for context selection, and use either cross-sectional designs or fairly commonplace measurement procedures. My research assesses the causal link between military service and self-image development with a more rigorous self-esteem measure and an analysis mode that yields precise estimates, accompanied by necessary controls.

Bronfenbrenner's (1979) ecological view of human development suggests that Vietnam-era military service should have an overall negative effect on self-esteem, when compared to those who took other paths. As the public mood turned against the war and eventually the young servicemen themselves (an exosystem phenomenon), many of those who served during this era often felt the sting of indifference, disapproval, and occasional hostility (see Figley and Leventman, 1980). Those on other paths experienced young adulthood and the sociohistorical events in a decidedly different light (a mesosystem phenomenon).

## *College*

College also presents a powerful institution of socialization (Bowen, 1977; Freedman, 1979; Newcomb, 1979). While college environments, curricula, and student body characteristics and interests range widely among institutions (e.g., Feldman and Newcomb, 1973; Panel on Youth, 1974, p. 87), general agreement reigns that higher education's basic goal is to transmit society's dominant knowledge and values to succeeding generations (Bowen, 1977; Kaysen, 1969; Ladd and Lipset, 1975). Bowen (1977, p. 8) identifies three functions: "knowing and interpreting the known (scholarship and criticism), discovering the new (research and related activities), and bringing about desired change in the cognitive and affective traits and characteristics of human beings (education)." Funk and Willits (1987) summarize college's impact on the individual:

> Both in and out of the classroom, the college student is likely to be exposed to new ideas, new people, and new environments, which would be expected to challenge his or her attitudes, values, and perceptions. Many believe that college "frees" the mind and provides opportunities for intellectual growth and personal development through contact with the diversity of ideas and persons . . . (p. 224).

Research on college's effects on personality development is perennial, going back at least to Newcomb's classic study of the liberalizing effect of faculty and college reference groups on the political attitudes of Bennington College students (Newcomb, 1943, 1965). Numerous other studies also find a general liberalizing trend in social and political attitudes among college students (e.g., Bowen, 1977; Astin, 1977; Feldman and Newcomb, 1973; and Funk and Willits, 1987).

Astin (1977, p. 189) argues that changes during college are often quite enduring, but that college experience itself may not be the power that instills strong values. It may be the "channeling" that occurs after college. He holds that college's liberalizing impact may extend over many years because college affords career paths and living patterns largely unavailable, and possibly quite foreign, to the typical high school graduate (p. 189). Bowen (1977, pp. 88ff) argues that an enduring effect of college is the cognitive development that undergirds lifelong learning.

Most studies support that college beneficially affects personality development (e.g., Astin, 1977, p. 34; Bachman, O'Malley, and Johnston, 1978, pp. 99–100; Bowen, 1977, p. 228; Feldman and Newcomb, 1973, p. 311). Astin (1977, p. 32–34) finds that among 25,000 students who entered college in 1966, and remained until 1970, had self-concepts that varied somewhat throughout the college years, but college attendance nevertheless increased one's sense of competence and self-worth. Greater education offers the opportunity to acquire a higher status occupation and income (Sewell and Hauser, 1975). High status jobs, moreover, tend to offer more challenge, autonomy, and satisfaction than lower level jobs. Such jobs thus tend to positively impact self-concept (e.g., Kohn and Schooler, 1983; Mortimer and Lorence, 1979) and could sustain positive self-concepts developed in college. Bowen (1977, p. 433) and others claim college's largest psychological impact is affective rather than cognitive.

College effects on self-concept, however, have not been found consistently positive. Using NLS-72 data, Kanouse and associates (1980) find that the only post-high school track that significantly affected self-confidence was four-year college, with an overall *negative* impact (p. 39). Although this runs counter to most evidence, plausibly some students did decline in self-worth during college. Since the college-bound group tended to have very high 12th-grade self-worth scores, later scores might dip somewhat once a more competitive milieu obtained. This argument echoes Davis's (1968) "frog pond" effect whereupon high ability students in selective colleges struggle for grades and recognition judged by local college standards rather than national standards (see also Rosenberg, 1981). Probably the most important reason for Kanouse and associates' findings is that all track classifications were based on subjects' *initial* post-high school track. The "college" track, for example, includes people who went to college directly after high school but may have subsequently dropped out to enter another track. Consequently, a track's effects are confounded with selection processes, duration effects, and other tracks' effects. Compared to other contexts, I would expect college to have if not a positive impact, then at least a neutral one. This seems particularly true since (1) self-esteem tends to resist change (see Wylie, Miller, Cowles, and Wilson, 1979), (2) the college-bound tend to have high prior self-esteem, and (3) college probably affects political and social values more than self-esteem (Alwin, Cohen, and Newcomb, 1991).

## *Work*

Marx's seminal materialist thesis argues that the nature of people's work and the objective conditions they live under generate their consciousness, feelings of alienation, and other central personality features (1964, pp. 122ff). More recently, Kohn and Schooler (1983) and Van Maanen (1976) provide useful insights into the relationship between work and personality. Van Maanen (1976) says that since nearly 90% of the American labor force works in an organizational setting (p. 67), workplace socialization is organizational socialization, the "process by which a person learns the values, norms and required behaviors which permit him to participate as a member of the organization" (p. 67). Because they are young and fairly recent full-time labor force entrants, workers in the Youth in Transition study might be particularly vulnerable to workplace pressure. It has been widely shown that poorer work conditions (especially in occupational self-direction and autonomy), reduces self-esteem and feelings of competence (e.g., Kohn, 1969; Kohn and Schooler, 1973, 1983; Mortimer and Lorence, 1979).

Agreement is widespread that organizational socialization is most intense and influential when a worker first enters an organization or moves across organizational boundaries (e.g., Kanter, 1977; Van Maanen, 1976, pp. 78–79; Van Maanen and Schein, 1979; Mortimer and Simmons, 1978). However, studies of psychological development during the transition to adulthood largely ignore people who take the traditional path (Hogan, 1981) of directly entering full-time labor after high school. Research focuses on the brighter, wealthier, and more upwardly mobile (the college-bound) and, to a lesser degree, those whose transition to adulthood is difficult. Studies of social stratification typically find that the lower one's social class, the lower one's social and economic participation and self-esteem (Foner, 1978, pp. 228–229; Rosenberg and Pearlin, 1978). People between ages 20 and 24 (and over 65) also earn the least (Foner, 1978, p. 224). Since the majority of men who directly enter work end their formal educations after high school, come from low SES backgrounds, have lower intellectual ability, and are new job incumbents, they generally occupy unskilled or manual jobs (Bachman, O'Malley, and Johnston, 1978), which typically offer little autonomy and control.

According to Borman (1988, p. 53), employers view young job applicants through two negative stereotypes. First, they believe young workers are either unwilling to work or "lack understanding of the demands of the workplace" (p. 53). Second, they view graduating high school seniors as labor market transients "'who will remain on the job only long enough to attain short term goals'" (Shapiro, 1983, p. 40 quoted in Borman, 1988, p. 53). In 1969, when most boys in the sample graduated from high school, those seeking civilian work aroused a third employer fear: they could be drafted, meaning a net loss for the company in employee training and development. Until the draft ended in 1972, this new fear discouraged hiring draft age men for the better, self-enhancing jobs—work that invests company time or dollars in employee development.

Sheppard (1973, pp. 99–100) detects a general lack of career planning in our society, especially among prospective and recent labor force entrants:

> This pattern involves—at least for working-class youth—haphazard choices of early jobs based on limited knowledge of the full range of types of occupations and kinds of employers, often motivated by a desire for independence from the family or a desire for money, even at low wages, in order to keep up with one's peers.

Many working class youths take intrinsically dissatisfying jobs with the "belief that they will escape them in time but find themselves, after the years ebb by, more or less 'trapped'"(p. 100).

These observations cause us to expect net reduced self-esteem among boys who directly enter the labor force after high school compared to those on other paths to adulthood. Quite limited data, however, support this expectation.

## Data and Methods

### *Specifications of the Structural Model*

The structural equation model was estimated using LISREL 7 (Joreskog and Sorbom, 1989) and consists of 20 exogenous variables (N = 1,090; 30% loss through missing data). Listwise deletion of cases avoids unstable parameter estimates caused by pairwise deletion (see Bollen, 1989, pp. 370–73). LISREL's analysis of linear structural relationships through maximum likelihood estimation allows one to: (1) estimate a latent measurement model while simultaneously estimating a structural equation model of the causal relationships among a set of exogenous and endogenous variables; (2) incorporate random and systematic measurement error, separating "real" change from change due to measurement error, which is crucial for accurate estimations of stability and causal influence; and (3) test overall model goodness-of-fit (GOF) after imposing theoretically derived constraints. GOF helps determine whether a given theoretical model adequately describes the pattern of relationships within a given data set (i.e., the variance-covariance input matrix). I report the goodness of fit index (GFI) (Wheaton, 1987) and the chi-square-to-the-degrees-of-freedom-ratio (the "fit ratio") (see Wheaton, Muthen, Alwin, and Summers, 1977; Carmines and McIver, 1981; Bollen, 1989).

The structural equation model has six components: (1) a selection bias term, (2) background variables, (3) context predictor variables, (4) Wave 3 self-worth, (5) the three post-high school social contexts and a time-in-context estimate, and (6) Wave 5 self-worth. The four dependent variables are the work, military, and college contexts, plus "other" for all unclassified subjects. (The "other" group eliminates the possibility of sample selection bias [see Berk, 1983; Maddala, 1983]). The 25 independent variables, measured during the first three waves (grades 10 to 12), comprise both social and psychological measures.

### *Selection Bias Correction Term*

The exogenous selection bias correction term[3] estimates through probit analysis the probability of being present in the Wave 5 (terminal) sample. Since attrition can seriously affect generalizability through biased parameters, selection bias control is crucial (Berk, 1983, Maddala, 1983). If the term has significant effects on other variables, then one may assume that selection bias has been detected and effectively neutralized; nonsignificance indicates the determinants of sample selection, as expressed in the hazard rate, have no bearing on the self-worth constructs.

*Background Variables*

The exogenous background variables are intellectual ability and family of origin socioeconomic status (SES). While both are significant context choice predictors, I depict them separately to underscore their importance in estimating the structural model of self-worth development. SES is widely believed to influence personality, including self-esteem (e.g., Rosenberg, 1965, 1979; House, 1981; Clausen, 1991; Kohn, 1969, 1977; Kohn and Schooler, 1983). Measured ability[4] is believed to be an important contributor to competency-based self-esteem, particularly as expressed in the impact that efficacious performance has on self-esteem (see Clausen, 1991; Gecas, 1982; Gecas and Schwalbe, 1983; Rosenberg, 1979; Rosenberg and Pearlin, 1978; Bandura, 1982). Competency-based self-esteem also comports with self-esteem formation and maintenance via the principles of self-attribution and social comparisons (see Rosenberg, 1979; Rosenberg and Pearlin, 1978; Gecas, 1982; Gecas and Schwalbe, 1983).

*Context Predictor Variables*

The exogenous context predictor variables control the determinants of context choice so that observed relationships between context and Time 5 self-worth are not spurious or biased. If students that are more confident disproportionately go to college, for example, any association between the educational context and self-esteem could derive from this selection process.[5] The predictor variables also place the individual within a larger field of relations (Dannefer, 1984; Bronfenbrenner, 1979). The classification schema, theoretical rationale, and literature review for the choice analyses are reported in detail elsewhere (Owens, 1992).

I specify a range of individual-level and sociocultural factors hypothesized to bear directly upon the selection of work, military, or college as a primary post-high school social context. Briefly, the context choice analyses are grouped under a conceptual framework involving six underlying choice domains and the context predictor variables subsumed under them (see Table 3.1). Significant predictors are starred.

The boys who chose work or the military as their main post-high school social context share some characteristics. Those classified in the college or "other" social contexts also share characteristics. Overall, the work context was chosen by boys from large families of the lower socioeconomic strata, who were in high school vocational tracks, and who worked more hours in their senior year, as compared to the college group. The workers tended to have the lowest intellectual ability and have friends who were least impressed by going to college of any context. Boys who early on believed that their parents wanted them to enter the work force after high school were significantly more likely to enter the work context than military context.

Similarly, the military context was chosen by boys with little desire to attend college. These boys generally came from large families in the lower socioeconomic strata, who followed a vocational track in high school, and who had poorer grades in contrast to the college group. The military-bound, as opposed to any other context, were most likely to have nonfarm backgrounds, a grade failure, and parents who would be happy if they served in the military. They tended to be more hawkish than the work group, but not significantly more than the college-bound.

The college context was chosen by boys with the smallest families and highest socioeconomic status. They also had the highest GPAs, followed a college track in high school, had a strong intention of going to college, and worked the least amount of hours in 12th-grade. Boys with higher intellectual ability and friends impressed by going to college were more likely to go to college than work.

*Social Context Variables*

The exogenous social context variables[6] are work, military, and college. Since context is a categorical variable, a dummy coding scheme assigns a man a 1 if he fits a context's inclusion criteria, otherwise a 0. Anyone not categorized into one of the three main contexts occupies a dummy residual group called "other" ($N = 548$), the omitted (or baseline) group for comparison. The work context comprises persons ($N = 183$) (a) in the full-time labor force during the years since high school (at least Wave 4, 1970, and Wave 5, 1974), (b) with less than one year of post-secondary schooling, and (c) never in the military. The college context ($N = 455$) requires at least four years of college or a baccalaureate and no military service. The military context ($N = 239$) indicates 18 or more months of active military duty, except for the handful of Vietnam veterans who served fewer.

Standardized total number of months in a particular context represents each respondent's mutually exclusive *time-in-context score*.

(Standardization is indicated because the possible length of time is greatest in the work context, as some left school and began work very early, and least in the military.)[7] Due to the original coding protocol, length of time in the work context reflects one's current job. This accords well with the need to be sensitive to the power of organizational socialization on personality (see above discussion).

*Endogenous Variable*

Six indicators measure the two latent self-worth constructs. Measurement parameters are freely estimated and incorporate correlated error terms between the same indicators measured across time—a correction for unreliability due to systematic measurement error (Cohen and Cohen, 1983; Long, 1983). Earlier confirmatory factor analyses support a two-factor self-esteem construct composed of positive and negative self-esteem. I focus on the positive dimension, dubbed self-worth. The latent Wave 3 self-worth variable is included as an important determinant of Wave 5 self-worth. Logit analysis shows that Wave 3 self-worth does not significantly predict any social context (Owens, 1992).

# Findings[8]

Figure 4.1 presents the maximum likelihood estimates. Appendix B contains the variance-covariance matrix and Appendix C the phi matrix of intercorrelations among exogenous variables.

Figure 4.1 shows a good fit between the model and the data (GFI = .963, fit ratio = 2.5). The military context significantly predicts Wave 5 self-worth (beta = $-.1$, p < .05), while the work context is slightly below the conventional cut-off (beta = $-.078$, p < .1). The college effect is not significant. Four of the 16 control variables significantly affect Wave 5 self-worth: intellectual ability and Wave 3 self-worth (p < .001), and number of siblings and time-in-context (p < .01).

## The Social Contexts

The military context has the greatest impact on early adult self-confidence, then work, and lastly college. The military path's size and

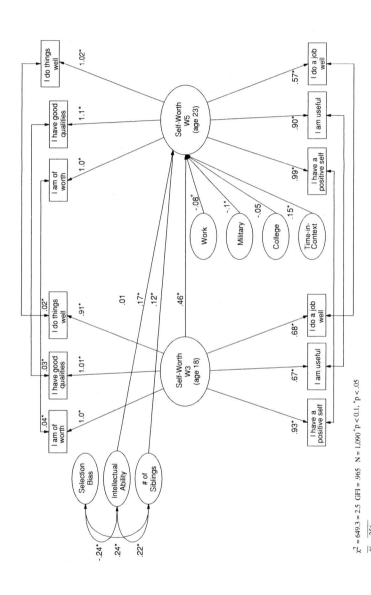

$\chi^2 = 649.3 = 2.5$  GFI = .965  N = 1,090 $^+$p < 0.1. $^*$p < .05

df  256

*Note*: Nonsignificant exogenous variable not shown: (1) family of origin socioeconomic level (Wave 1), (2) first born (Wave 1), (3) reared on farm (Wave 1), (4) folks wanted S to go to work after high school (Wave 1), (5) folks would be happy if S served in military after high school (Wave 1), (6) GPA in 10th Grade (Wave 1), (7) S in college high school track (Wave 3), (8) S failed a grade before 9th (Wave 1), (10) Hawkish war attitude (Wave 3), (11) S aspires to college (Wave 3), (12) S aspires to tech. school (Wave 3), (13) hours worked in 12th grade (Wave 3).

**Figure 4.1.** The influence of post-high school social context on young adult self-worth (standardized beta and gamma coefficients).

sign shows the small, negative net impact predicted. While neither the military nor work explicitly seek to change the individual, each socializes new members to their roles and the organization. While the military may aim more to control than fundamentally change the individual, many youths enter expecting some measure of social recognition, approval, and personal growth. I argue that many veterans of this era, instead, experienced an acute twist to the veteran's plight as depicted in Kipling's (1940) famous poem Tommy[9], leading to a small but significantly negative impact on their self-esteem.

Length of time in one's particular context emerges as a key social context variable. As time-in-context increases, so does self-worth (beta = .147, p < .01). This finding comports well with self-efficacy theory (e.g., Bandura, 1977; Gecas, 1989), wherein the expectation that one can perform successfully and also exercise some influence or control over the environment are key to feeling efficacious and competent (Gecas, 1989, p. 294). By extension, the longer an individual participates in a specific social context, and the more acclimated one should become, the more competent and efficacious, and self-confident one should feel. Conversely, the powerful self-esteem motive—wherein one seeks to both protect and enhance self-esteem (see Rosenberg, 1979, pp. 54–57)—accords highly with the notion that those experiencing difficulty in a context would hope to hasten their departure.

My theoretical schema helps put the social context findings into perspective. While Elder (1974, 1986, 1987) and Elder and Clipp (1989) note that service during World War II and the Korean War sometimes had long-lasting psychological effects, both positive and negative, they also emphasize contextual effects such as social and historical circumstances (e.g., war context and homecoming experiences) (Elder and Clipp, 1989). Recalling the sociogenic thesis' entreaty to consider the larger societal forces impacting the developing individual (Dannefer, 1984) and Bronfenbrenner's (1979) ecological perspective, I note in reference to the mesosystem (i.e., relations between contexts) that some of the well documented social disapproval the military group in this era experienced (see Figley and Leventman, 1980 for a thorough discussion) likely flowed from their counterparts who chose work and college over the military. Because of both reflected appraisals (Sullivan, 1953; Cooley, 1922) and social comparisons (see Rosenberg, 1979, pp. 62–70), one could expect that the general castigation many veterans felt (stemming from the unique social stigma emanating from the larger social and historical processes operating) would lower self-worth. In this sense, the social and political environment (i.e., the exosystem) touched the military group in a qualitatively different manner than their nonmilitary peers. The negative association between the military context and self-worth seems all the more compelling in that, except for the time-in-context measure, the model includes only gross context effects. Therefore, even without knowing such microsystem variables as combat exposure or military occupation, we see the military's generally deleterious psychological effect.

Historical context does not solely account for the deleterious effect of the military (and possibly work). And even though the work effect lacks significance at the .05 level, the results suggest, in relation to the other contexts, that work too may somewhat negatively affect self-worth, especially in reference to organizational position. The work and military contexts—especially for people in lower level positions of each hierarchy—probably share similar general features. Recall, for example, Van Maanen's (1976) observation that 90% of all jobs occur in an organizational setting, which connotes a hierarchy or bureaucracy in nearly every workplace. And the military, of course, is a huge bureaucracy. Even without detailed job information about the workers and the servicemen, one may assume that youthfulness, a general lack of post-high school training or education, and relatively recent movement into his organization, placed most study participants in low level unskilled or semi-skilled manual or technical positions.[10] Such jobs typically offer little autonomy, low occupational self-direction, and meager cognitive complexity, all of which link negatively to psychological well-being (e.g., Bandura, 1997; Sheppard, 1973; Gecas, 1981; Kohn, 1969, 1977; Kohn and Schooler, 1983; Mortimer and Lorence, 1979). Gunderson (1976, p. 70), for example, finds that Navy enlisted men in unskilled and manual positions generally show higher rates of reported psychiatric disorders than men in more highly skilled positions.

College is not significantly related to young adult self-worth. Although one might have expected a small positive correlation, the college-bound have fairly high prior self-esteem scores (mean = 3.98 on a 5-point scale), leaving little room for improvement in a construct that tends to generally resist change (Wylie et al., 1979; Wells and Marwell, 1976). Furthermore, the college experience arguably differs largely from the military or even work. While the larger sociohistorical processes operating may very well have impacted students' social and political attitudes, no compelling reason requires, as for the military group, that they would directly influence their feelings of worth. As noted earlier, while the literature strongly indicates that college may powerfully affect personality (e.g., Funk and Willits, 1987), its strongest impact is generally on social and political values, rather than self-esteem.

## *Other Variables*

As might be expected, Wave 3 self-worth has the strongest net effect on Wave 5 self-worth (beta = .459, p < .001). This coefficient shows the degree of normative stability[11] (Mortimer, Finch, and Kumka, 1982, p. 276) in self-worth between grade 12 and five years after high school. The stability coefficient is not particularly large, implying subjects' personalities were shifting in the post-high school years, when most were confronting adulthood. Still, some moderate stability is also indicated. (See also O'Malley and Bachman [1983] on the general stability of self-esteem between adolescence and early adulthood, and Mortimer, Finch, and Kumka, 1982.) This persistence in self-concept accords with the self-consistency motive (Rosenberg, 1979, p. 57), and helps explain the modest context effect.

Intellectual ability has the next strongest effect on young adult self-confidence (beta = 0.174, p < .001). Rosenberg's (1979) principles of social comparisons and self-attribution (pp. 67–73) might well account for this relationship. In social comparisons, people learn about themselves by comparing themselves to others, which leads to favorable or unfavorable self-ratings (pp. 67–68). If early adulthood is a period of training, initial career entry, and adoption of important new social roles (Levinson, 1978; Mortimer and Simmons, 1978; Borman, 1988)—and if intelligence is associated with academic achievement, ambitions, and skills (e.g., Spenner and Featherman, 1978)—then the intellectually ablest are probably more successful at school and work, and have more self-enhancing military jobs. Through social comparisons, those who are more successful in relation to their peers feel more positively toward themselves (see also Davis, 1966).

Somewhat surprising is the effect number of siblings has on Wave 5 self-confidence (beta = .116, p < .01). (Sibship is included as a significant context choice predictor.) All things being equal, the data indicate that young adult self-worth increases with sibship. This seems to contradict research on the relationship between adolescent self-esteem and sibship. The oldest perspective concerning the effect of family size on children is the "dilution model" (Heer, 1985):

> [a]n increase in the number of siblings or a decrease in the spacing between them dilutes the time and the material resources that parents can give to each child and that these resource dilutions hinder the outcome for each child (p. 28).

As sibship increases, intellectual ability (e.g., Heer, 1985, pp. 31–39; Clausen and Clausen, 1973, p. 195; Blake, 1989), educational attainment (Heer, 1985; Blake, 1989), and educational aspirations (Blake, 1981) generally decline. However, the effect of family size on personality development is less clear. Clausen and Clausen (1973) suggest that if larger families provide less parental support and interaction, they might also yield greater support and "mentoring" from other siblings (p., 199). Less parental support may compel greater resourcefulness in obtaining alternative support. Clausen and Clausen (1973) write that while Rosenberg (1965) finds teenage self-esteem tends to be higher among only children, family size effect on self-esteem is inconsistent in families of two or more children (p. 199). Moreover, when background characteristics are controlled, the relationship between self-esteem and sibship has been found to disappear (Ernst and Angst, 1983, p. 80). Few studies explore the self-esteem and sibship relationship among adults (p. 80).

It seems reasonable to speculate that larger families may provide a young adult with both a larger network of close personal contacts for support and an environment that may foster more personal resourcefulness, independence from one's parents, and adaptability to others' demands and personalities. They may also occasion role modeling for or emulation of other siblings. Larger families may thus benefit personality development during the transition to adulthood, especially as it concerns self-worth.

Finally, two variables not found to be significant are worth noting: the selection bias term and SES. Even though 23% of Wave 1 whites attrited by Wave 5, attrition-based selection bias does not seem to be a serious problem in the estimation of self-esteem development. Of further interest, that SES is not significant seems to contradict other research on the relationship between SES background and self-concept (e.g., Kohn, 1969, 1983; Gecas, 1979, 1981). However, the literature also suggests that SES is often not a very salient factor in an adolescent or young adult's self-concept (Rosenberg and Pearlin, 1978). Other, more immediately identifiable bases of social comparisons (e.g., Gecas and Schwalbe, 1983, p. 83; Rosenberg, 1979, p. 68), such as physical appearance, grades, talents, and so forth, may be of greater importance.

## Summary and Conclusions

This chapter examines the impact of three post-high school social contexts—full-time labor force, the federal military, and college—
on the positive dimension of self-esteem called self-worth, using the
sociogenic thesis as posed by Dannefer (1984), Glenn's (1980) aging-
stability hypothesis, and Bronfenbrenner's (1979) ecological theory of human development. Briefly, the sociogenic thesis urges human development
researchers to eschew a narrowly defined age-graded view by incorporating a sociological component sensitive to the different social environments that influence individuals over the life course. This reason
compels incorporating the three social contexts of work, military, and
college in a model of self-esteem development. However, focus on intra-
context processes (e.g., combat intensity, college grades, occupational
self-direction) would preclude side-by-side comparison of context effects within a single causal model. My model concentrates instead on
inter-context effects, incorporating selection bias, background variables,
context predictors, prior self-worth, the three social contexts, and early
adult self-worth (i.e., positive self-esteem). Although intra-contextual processes are important, and subsequent research might fruitfully exploit
this issue, the present research is the first I have seen which examines
gross context effects through a model incorporating variables held by
all.

Of the three contexts, the military has the most significant (negative) net impact on self-concept during the transition to adulthood,
followed by work (slightly negative), and college (no impact). The sociogenic thesis and Bronfenbrenner's ecological perspective roots the observed negative effect of the military, and possibly work, in the larger
sociohistorical processes impinging on the individual, and on the workers' and servicemen's low organizational positions. In contrast to the military or work after high school, college does not appear to appreciably
affect self-esteem. This may be due to the college-bounds' relatively high
prior self-esteems and because college may impact social and political values more than self-worth. Unlike previous studies, I found time-in-context
to be an important (positive) predictor of later self-worth. Since time-in-
context likely operates through one's acclimation to role demands, time
should enhance feelings of mastery and control, and consequently self-
esteem.

# Notes

1. Kanouse and associates (1980) speak of "self-esteem," not self-worth. However, since composed of four of Rosenberg's positive self-esteem items, their construct is called self-worth here.
2. Bachman, O'Malley, and Johnston (1978) do find that boys in the military eventually had lower self-esteem in 10th grade, but by the end of high school theirs matched the others'.
3. See Chapter 2 for a discussion of the Probit analyses and the rationale for estimating the selection bias term.
4. See Chapter 2 for a full description of the intellectual ability construct.
5. Thus, inclusion of context predictor variables attempts methodologically to avoid potentially serious omitted variable bias (Maddala, 1983). For example, bias would exist in the path between social context and T5 self-confidence (path c) were the context predictor causally related to both social context (path a) and T5 self-worth (path b) (see Figure 4.2).

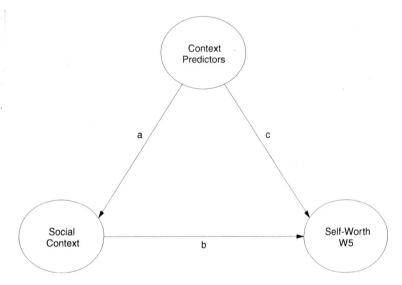

**Figure 4.2.** Omitted variable bias.

Context choice predictors reduce the likelihood of such misspecification, and help estimate more bias-free parameters on the effect of social context on self-worth. The choice schema is tested with multinomial logistic regression (MLR) (see Owens, 1992), a generalized linear modeling technique appropriate with a multiple categorical dependent variable and continuous and categorical independent variables (see Weisberg, 1985, pp. 260, 267–271; Aldrich and Nelson, 1984, pp. 30ff ). LIMDEP 8.0 (Greene, 2002) is used to obtain the maximum likelihood coefficients.

6. Although prior logit analyses indicate several significant context predictors, the contexts must be treated as exogenous rather than endogenous variables because dummy endogenous variables are highly problematic in causal models. The central difficulty is that because dummy variables have skewed distributions, restricted variances, and non-normally distributed error terms, they produce poor parameter estimates as endogenous variables. All such causal relations are expressed in the phi coefficients.

7. No significant interaction between social contexts and time-in-context obtains when the interactions terms were included in the causal models (following Bollen's, 1989, pp. 129 guidance). A separate analysis of multicollinearity (following the procedures Berry and Feldman, 1985, pp. 42–46 outline) indicates it too is not a significant problem.

8. A model with self-deprecation (i.e., negative self-esteem) shows that besides Wave 3 self-deprecation (beta = .421, p < .001), the only other significant relationship to Wave 5 self-deprecation is perceived parents' preference for their son to work after high school, as measured in 10th-grade (beta = .093, p < .05).

9. ... You talk o' better food for us, an' schools, an' fires, an' all: We'll wait for extra rations if you treat us rational.
   Don't mess about the cook-room slops, but prove it to our face
   The Widow's Uniform is not the soldier-man's disgrace.
   For it's Tommy this, an' Tommy that, an' "Chuck him out, the brute!"
   But it's "Saviour of 'is country" when the guns begin to shoot;
   An' it's Tommy this an' Tommy that, an' anything you please;
   An' Tommy ain't a bloomin' fool—you bet that Tommy sees!

10. The vast majority in the military were low ranking enlisted men; over three-fourths (77%) reported a high rank of E-4 (equivalent to a corporal) or E-3 (equivalent to a private first class). Persons with the rank of E-3 or E-4 typically have little or no supervisory responsibility or organizational power and often occupy the lowest technical, clerical, or unskilled manual positions. Based upon their civilian Duncan SEI scores for Wave 5 jobs, the veterans had the lowest status jobs (mean = 27.4), followed by the workers (31.9), and the college group (49.8). Jobs typical of these scores are boilermaker and brickmason apprentice (workers), farm implement mechanic and household appliance and accessory installer (veterans), and salesman and bookkeeper (college group) (Hauser and Featherman, 1977). Due to extensive missing data, however, I exclude this variable from the structural model.

11. Normative stability is the persistence of individual ranks or differences in an attitude over time (Mortimer, Finch, and Kumka, 1982, p. 267). The moderately positive correlation between the repeated measures of self-esteem (b = .459) indicates general stability in the five year span; that is, individuals who score high or low on self-esteem in 12th-grade, in relation to others, tend to retain the same relative positions in early adulthood. Using Monitoring the Future data, O'Malley and Bachman (1983) also show that while self-esteem tends to have fairly high normative stability—especially year-by-year—its mean level stability (i.e., an attitude's overall quantity, strength, or magnitude over time) tends to change during adolescence and early adulthood as group self-esteem scores slowly increase (O'Malley and Bachman, 1983, pp. 264–264).

# 5 —————————————————————————

# Summary and Conclusions

## Summary of the Literature

The impact of work experience on psychological development has been well established in the sociological literature on social structure and personality, going back at least to Marx's (1964) early writings and extending to the highly regarded research of Kohn and his colleagues (Kohn et al. 2002; Kohn et al. 2000; Kohn and Schooler, 1983; Kohn, 1969, 1977). The effect of higher education (e.g., Newcomb, 1943; Astin, 1977; Bowen, 1977; Sanford and Axelrod, 1979) and military service (e.g., Moskos, 1970; Borus, 1976; Egendorf and associates, 1981; Elder, 1986) have also been given attention. The present study of the impact of post-high school social context on self-esteem is an outgrowth of this research tradition in the area of social structure and personality and life course analysis.

While most life course theorists agree that adolescence and young adulthood are important years in the development of adult attitudes and identity, chapter 1 showed that a growing body of sociological research points to the important role that social environments and organizations play in the development of adult personality as well. The research to date, however, has not been brought together in a comprehensive analysis of the effect of post-high school social context on personality development. Little attention has been directed to how the three major social contexts persons typically enter after the completion of secondary school influence their development during the transition to adulthood. This study was designed to help bridge the gaps in the literature by examining how participation in the full-time labor force, the active Federal military, and college impact

153

self-esteem during the transition to adulthood. The four objectives of this study were:

1. To examine the empirical dimensionality of the self-esteem construct, through confirmatory factor analysis;
2. To correct the sample for selection bias due to attrition over the eight year span of the study, through probit modeling;
3. To develop a context choice schema which predicts entrance into one of the post-high school social contexts, through multinomial logistic regression;
4. To examine the role that post-high school social contexts play in psychological development during the transition to adulthood, through the use of structural equation modeling.

The literature review in chapter 1 showed that while stage theories of the life cycle (Erikson, 1963, 1982; Levinson, 1978; Sears, 1981) have been accepted explicitly or implicitly by many life course researchers, stage theories often fall victim to an ontogenetic fallacy (Dannefer, 1984). This fallacy stems from an overemphasis on age-graded (i.e., stage-specific) change which too often fails to acknowledge the important role that diverse social environments play in human development, especially in adulthood (Bandura, 1997; Dannefer, 1984; Elder, Johnson, and Crosnoe, 2003). Three key life course perspectives were introduced in chapter 1 which incorporate a strong sociological view of human development and which offer alternatives to ontogenetic explanations of the life course: Dannefer's sociogenic thesis, Glenn's aging-stability hypothesis, and Elder, Johnson, and Crosnoe's paradigmatic principles of life course theory.

The sociogenic thesis emphasizes the importance of environmental influences on human development, and the situational diversity of these influences. It along with the life course principles informed the analyses presented in chapter 4, dealing with the effects of post-high school social contexts. The sociogenic thesis also features the active role of the individual in choosing and molding his or her environment, represented in this study by the context choice schema presented in the logit analyses of chapter 3. The sociogenic thesis and Elder and associates' life course principles form the backbone of my investigation of the role that post-high school social context plays in the development of self-esteem.

The aging-stability hypothesis posed by Glenn (1980) complements Dannefer's sociogenic thesis and Elder, Johnson, and Crosnoe's life course principles. As discussed in chapter 1, the aging-stability hypothesis holds that "attitudes, values, and beliefs tend to stabilize and to become less likely to change as a person grows older" (Glenn 1980, p. 602). Glenn (1980) discusses the aging-stability hypothesis in terms of developmental and environmental influences, but emphasizes the latter by asserting that the environment may be the critical factor in the observed decrease in "change-proneness" over the life course. Attitude stabilization over the life course was considered to be largely due to a general consolidation of social experiences and social roles as people move through adulthood. The adult development literature generally supports the idea that adults are most susceptible to change when they enter new organizational contexts, move across organizational boundaries, or when salient social roles are acquired or relinquished (e.g., Lutfey and Mortimer, 2003; Gecas, 2000; Bush and Simmons, 1981; Moss and Sussman, 1980; Mortimer and Simmons, 1978; Van Maanen, 1976, 1979).

The literature on socialization to roles links socialization to psychological change. This position is contrary both to the idea that aging itself tends to stabilize attitudes, and to the developmental paradigm which confines psychological development to specific age periods. Glenn criticizes the idea that human development largely ceases when adulthood is attained. Since the period of transition between adolescence and adulthood is a time when the roles of childhood and adolescence are relinquished and new adult social roles and organizational affiliations are acquired, this period was selected as my research focus.

The need to assess the aging-stability hypothesis through attitudes with stable or highly abstract objects (Glenn 1980, p. 605) was emphasized in chapter 1. If attitudes with low ego involvement and low salience for the individual are considered, then rapidly shifting attitudes may be mistakenly seen as indicating phases of development (Sears, 1984). Since the self is indeed an abstract and relatively stable object (e.g., Owens, 2003; Rosenberg, 1978; Wylie, 1974; Wells and Marwell, 1976), self-esteem was selected as the main focus of this study. However, because self-esteem is a theoretical construct, it was necessary to establish its dimensionality and to test it for structural invariance. In chapter 2 it was shown that establishing structural invariance is critical to both the theoretical formulation of the self-esteem construct and to the assessment of its stability over time. Structural invariance exists when a construct is "characterized by the same dimensions, and when there is a persistent pattern of relations among its component attributes, over time" (Mortimer Finch, and Kumka, 1982, p. 266).

The process of occupational choice was also considered in chapter 1. A central theme emergent from the literature is that occupational choice is a multidimensional phenomenon that is part of a larger process. The focus of this study is not on occupational choice per se, but on the preliminary choice of the post-high school social contexts of work, the military, and college, each of which may heavily constrain occupational choice and attainments in the future. Prior conceptualizations of the occupational choice process (especially the insights of Blau and associates, 1956) were used to frame the empirical analyses of post-high school context choice presented in chapter 3.

## Summary of the Findings and Research Objectives

Bachman's Youth in Transition (YIT) study (Bachman, O'Malley, and Johnston, 1978), conducted at the University of Michigan between 1966 and 1974, provided the data for these analyses. As described in chapter 2, YIT is a nationally representative, five-wave longitudinal study of 2,213 tenth grade boys enrolled in over 80 American public high schools in the autumn of 1966. Data were collected in tenth grade (fall 1966), eleventh grade (spring 1968), twelfth grade (spring 1969), one year after high school (spring 1970), and five years after high school (spring 1974). This dissertation focused on self-esteem change between Wave 3 (1969) and Wave 5 (1974).

## *Research Objective 1: Assessing the Dimensionality of Self-Esteem*

The investigation of the empirical dimensionality of the self-esteem construct was presented in chapter 2. Determining the dimensionality of self-esteem is a key consideration in establishing its reliability and structural invariance. Rosenberg (1965) sees self-esteem as only one manifestation of self-conception, a "positive or negative attitude toward a particular object, namely, the self" (p. 30). However, while Rosenberg conceptualizes self-esteem as a unidimensional or global construct, other sociologists (e.g., Oates, 2004; Owens, 2000, 1994, 1993; Kohn and Schooler, 1983; Mortimer and Finch, 1983; Kohn, 1977) have separated his self-esteem construct into its positive and negative self-evaluative components.

Briefly, two measurement models were estimated using confirmatory factor analysis. The analyses were performed in order to determine whether the self-esteem construct measured in grade twelve (Wave 3) and five years after high school (Wave 5) was best characterized by a global self-esteem construct (i.e., a one-factor model) or a multidimensional construct (i.e., a two-factor model). The analyses showed that the data support a multidimensional self-esteem construct rather than a global construct. As noted, the multidimensional conceptualization of self-esteem consisted of a self-worth construct composed of six positively worded self-esteem items and a self-deprecation construct composed of four negatively worded items. (The possibility of a halo effect in the constructs was tested and found not to be operating.) Since my primary interest is in positive self-esteem, the analyses proceeded with the self-worth construct.

## Research Objective 2: Correcting the Sample for Selection Bias

Chapter 2 also investigated sample selection bias caused by possible nonrandom attrition from the Wave 1 sample. It was argued that selection bias may reduce the generalizability of longitudinal analyses when subjects are systematically included (or excluded) from subsequent waves. Without correcting for sample bias, the parameter estimates in the causal analyses can be biased, and the findings incorrect. In order to correct for selective attrition, a probit analysis was performed. A selection bias correction term, which expresses the probability of remaining in the Wave 5 sample, was calculated for each person. When the selection term was included as a regressor in the analyses presented in chapter 4, it was found to be a nonsignificant predictor of self-esteem.

## Research Objective 3: Identification of Context Choice Predictors

The primary goal of chapter 3 was to identify significant context choice predictors for later inclusion as controls in the causal analyses presented in chapter 5. Since a context choice component was implied in the chapter 4 causal models, failure to control for context choice might lead to model misspecification. Without controlling the predictors of context choice, the relationships between post-high school social context and later self-worth could be spurious.

The analyses examined the influence of 25 predictors of entrance into the work, military, or college contexts. Fifteen of the variables were found to be significant predictors. Briefly, the boys who chose work or the military as their main post-high school social contexts had somewhat similar characteristics, and those who chose college or were in the "other" (residual) group had similar characteristics. The work context was more likely to be chosen by boys from larger families and in the lower socioeconomic strata, as compared to those in the college and "other" group. The workers tended to have the lowest intellectual ability and the least hawkish war attitudes. The boys in the work context also tended to have a low desire to attend college and to have friends who were the least impressed by going to college.

The military context, like the work context, was disproportionately chosen by boys who expressed little desire to attend college and were from larger families and in the lower socioeconomic strata. The military-bound boys, as opposed to boys headed for one of the other contexts, were most likely to come from nonfarm backgrounds, to have failed a grade, and to believe their parents would be happy if they served in the military. The later two point to strong possible motivations to serve, one by proving something to and about yourself and the other to please one's parents.

The college context was chosen by boys who came from the smallest families and the highest socioeconomic status backgrounds. Not surprisingly, the boys with higher intellectual ability, the highest ninth grade GPA, and the strongest desire to go to college were more likely to choose college over competing contexts. Furthermore, the less a boy worked during twelfth grade, the more likely he was to go to college after high school; and, the more a boy's high school friends were impressed by going to college, the more likely a boy was to go to college and remain there. Peer influences certainly mattered in this case.

## *Research Objective 4: Effect of Post-High School Context on Self-Esteem*

This research objective was addressed by specifying two structural equation models designed to assess the influence of post-high school social context on psychological development, particularly as it pertains to the positive component of self-esteem, or self-worth. Dannefer's (1984) sociogenic thesis, Glenn's (1980) aging-stability hypothesis, and Elder, Johnson, and

Crosnoe's (2003) life course paradigm principles provided the theoretical rationale for examining the effect of social context on self-worth.

The results from the two models (a full model composed of 20 exogenous variables, and a reduced model composed of seven exogenous variables) provided modest support for the hypothesis that post-secondary social context influences self-esteem formation in early adulthood.

The military and work contexts significantly predicted Wave 5 self-worth, even though the work context was slightly below the conventional .05 cut-off. Four out of the 16 control variables were significant in predicting Wave 5 self-worth: intellectual ability, Wave 3 self-worth, number of siblings, and time-in-context.

The military context had a small but negative impact on early adult self-worth while the work context had a slight negative impact on those boys' feelings of self-worth. While neither the military nor work organizations explicitly seek to change the individual, each socializes new members to their roles and the organization. Since the military is probably more interested in control rather than fundamentally changing individuals, many youths enter the service expecting some measure of social recognition and approval, and personal growth. Unfortunately, many in the Vietnam-era felt that their service was not just unappreciated, but actually denigrated and stigmatized. This quite possibly led to the small but significantly negative impact the military had on the self.

Length of time in one's particular context also emerged as a key social context variable. Longer periods in-context increased self-worth. This finding corresponds well with self-efficacy theory. By extension, the longer one participated in a particular social context, and the more acclimated he probably became to it, the more competent and efficacious, and self-confident he likely felt. Conversely, the powerful self-esteem motive—wherein one seeks to both protect and enhance self-esteem—comports with the notion that those experiencing difficulty in a context would hope to hasten their departure.

My theoretical schema helped put the social context findings into perspective. The social and political environment touched the military group in a qualitatively different manner than their nonmilitary peers. The negative association between the military context and self-worth seems all the more compelling in that, except for the time-in-context measure, the model includes only gross context effects. Therefore, even without knowing such microsystem variables as combat exposure or military occupation, we saw the military's generally deleterious psychological effect.

Historical milieu does not account solely for the deleterious effect of the military (and possibly work) contexts. And even though the work effect lacked significance at the .05 level, the results suggest, in relation to the other contexts, that work too may somewhat negatively affect self-worth, especially in reference to organizational position. The work and military contexts—especially for people in lower levels positions of each hierarchy—probably share similar features. Even without detailed job information about the workers and the servicemen, one may assume that each person's youthfulness, general lack of post-high school training or education, and relatively recent movement into his organization, placed most in low level unskilled or semi-skilled manual or technical positions. Such jobs typically offer little autonomy, low occupational self-direction, and meager cognitive complexity, all of which link negatively to psychological well-being.

Although one might have expected a positive correlation between college and self-worth, the college-bound already had fairly high prior self-esteem scores, thus leaving little room for improvement in a construct that generally resists change. Furthermore, the college experience arguably is qualitatively different from the work context, but especially a total institution like the military. While the larger sociohistorical processes operating may very well have impacted students' social and political attitudes, no compelling reason emerged, as it did for the military group, that they would directly influence their feelings of worth. As noted earlier, while the literature strongly indicated that college may powerfully affect personality, its strongest impact is generally on social and political values, rather than self-esteem.

The effect of number of siblings on Wave 5 self-worth was somewhat of a surprise. The data indicated that young adult self-worth increased as sibship also increased. This seemed to contradict the literature. Increased sibship is usually associated with decreased intellectual ability, educational attainment, and educational aspirations. However, the effect of family size on personality development is less clear. If larger families provide less parental support and interaction, they might also yield greater support and "mentoring" from other siblings. Less parental support may also compel children to develop greater resourcefulness in obtaining alternative support. Even though teen-age self-esteem tends to be higher among only children, family size effect on self-esteem is inconsistent in families with two or more children.

It seems reasonable to speculate that larger families may provide a young adult with both a larger network of close personal contacts for support and an environment that may foster more personal resourcefulness, independence from one's parents, and adaptability to others' demands and personalities. They may also occasion role modeling for or emulation of other siblings. Larger families may thus benefit personality development during the transition to adulthood, especially as it concerns self-worth.

Finally, two nonsignificant variables are worth noting: the selection bias term and SES. Even though 23% of Wave 1 whites attrited by Wave 5, attrition-based selection bias did not seem to be a serious problem in the estimation of self-esteem development. SES's nonsignificance seemed to contradict other research on the relationship between SES background and self-concept. However, the literature also suggests that SES is often not a very salient factor in an adolescent or young adult's self-concept.

## Coda

Life course analyses must—and generally does—take an interdisciplinary perspective that is sensitive to both developmental and environmental issues. Although mixed support for the impact of social context on personality development was found in this study, the data set analyzed was not optimally designed to address the questions I posed. Specifically, the data did not allow assessment of within context variation in roles and job characteristics. Our understanding of how social contexts affect adult personality during the transition to adulthood could be furthered by designing studies that are sensitive to the wide range of environments and roles within the work, military, and college contexts. Moreover, if there is a lagged effect of social context on self-concept, there is a need to extend longitudinal studies begun in the adolescent years to middle and late adulthood. We must await future, intensive and comprehensive longitudinal studies before we can fully understand the implications of post-high school social context for the development of self-image during the transition to adulthood.

# Appendix A. Complete Multinomial Logistic Regression Model Predicting Entrance into a Particular Post-High School Social Context

| Independent variable | Military vs. work $b$ SE($b$) | College vs. work $b$ SE($b$) | Other vs. work $b$ SE($b$) | College vs. military $b$ SE($b$) | Other vs. military $b$ SE($b$) | Other vs. college $b$ SE($b$) |
|---|---|---|---|---|---|---|
| *Family Contingencies* | | | | | | |
| First born child (1 = oldest or only, 0 = otherwise) | -0.28 (.31) | -0.06 (.33) | 0.30 (.26) | 0.22 (.33) | 0.58* (.26) | 0.36 (.26) |
| Number of siblings | 0.06 (.07) | -0.34*** (.09) | -0.13* (.06) | -0.41*** (.09) | -0.19*** (.06) | 0.21** (.07) |
| Farm family origin (1 = from farm, 0 = otherwise) | -0.91* (.43) | 0.59 (.41) | 0.30 (.31) | 1.50*** (.48) | 1.21*** (.38) | -0.29 (.33) |
| Family socioeconomic status | 0.14 (.09) | 0.46*** (.09) | 0.32*** (.08) | 0.31*** (.09) | 0.17* (.07) | -0.14* (.06) |
| *Family and Peer Influences* | | | | | | |
| Folks want work after high school—10th-grade (1 = folks want work after high school, 0 = otherwise) | -0.90* (.45) | 0.22 (.51) | -0.18 (.33) | 0.65 (.56) | 0.72 (.41) | 0.07 (.44) |
| Folks want college after high school—10th-grade (1 = folks want college after H.S., 0 = otherwise) | 0.11 (.28) | 0.06 (.28) | 0.13 (.23) | -0.05 (.28) | 0.03 (.23) | 0.07 (.21) |
| Folks' happiness about entering military—12th-grade (−1 = unhappy, 0 = neutral, +1 = happy) | 0.51** (.20) | -0.42 (.23) | -0.15 (.18) | -0.93*** (.21) | -0.66*** (.16) | 0.27 (.18) |
| Peers' happiness about entering military—12th-grade (−1 = unhappy, 0 = neutral, +1 = happy) | 0.09 (.22) | 0.35 (.22) | 0.09 (.18) | 0.26 (.22) | -0.01 (.18) | -0.26 (.17) |
| Siblings' happiness about entering military—12th-grade (−1 = unhappy, 0 = neutral, +1 = happy) | -0.16 (.23) | -0.36 (.24) | -0.13 (.20) | -0.20 (.23) | 0.03 (.18) | 0.23 (.19) |
| Peers' respect for going to college—12th-grade (1 = not at all, 2 = somewhat, 3 = very true) | 0.57** (.20) | 0.74*** (.19) | 0.41** (.17) | 0.18 (.19) | -0.17 (.16) | -0.33* (.14) |

*School Performance and Experiences*

| | (1) | (2) | (3) | (4) | (5) |
|---|---|---|---|---|---|
| Grade point average in 9th grade | −0.005 (.02) | 0.08*** (.02) | 0.002 (.02) | 0.08*** (.02) | −0.07*** (.02) |
| College-bound high school track (1 = in college-bound track, 0 = otherwise) | −0.12 (.29) | 0.93** (.31) | −0.06 (.25) | 1.05*** (.31) | −0.98*** (.24) |
| Vocational-bound high school track (1 = in vocational-bound track, 0 = otherwise) | 0.10 (.38) | −0.37 (.48) | 0.17 (.32) | −0.47 (.49) | 0.54 (.41) |
| Grade failure before 9th grade (1 = failed grade(s), 0 = otherwise) | 1.81*** (.46) | 0.10 (.60) | 0.66 (.43) | −1.70*** (.55) | 0.56 (.50) |

*Attitudes Toward Self and Society*

| | (1) | (2) | (3) | (4) | (5) |
|---|---|---|---|---|---|
| Self-confidence—12th grade (1 = low,...,5 = high self-confidence) | −0.22 (.24) | −0.22 (.24) | −0.28 (.20) | −0.0001 (.24) | −0.05 (.18) |
| Ambitious work values scale—12th grade (1 = low ambition,..., 4 = high ambition) | −0.18 (.24) | −0.18 (.24) | −0.34 (.20) | −0.16 (.24) | −0.16 (.17) |
| Hawkish war attitudes scale—12th grade (1 = low hawk,...,5 = high hawk) | 0.73* (.36) | 0.45 (.30) | 0.69** (.25) | −0.28 (.24) | 0.24 (.22) |
| Dovish war attitudes scale—12th-grade (1 = low dove,...,5 = high dove) | 0.41 (.33) | −0.04 (.30) | 0.12 (.26) | −0.45 (.31) | 0.16 (.22) |
| Vietnam War dissent scale—12th-grade (1 = low dissent,...,5 = high dissent) | −0.01 (.32) | −0.08 (.30) | 0.25 (.26) | 0.10 (.30) | 0.16 (.21) |

*Ambitions and Planfulness*

| | (1) | (2) | (3) | (4) | (5) |
|---|---|---|---|---|---|
| Strength of college attending plans—12th-grade (1 = definitely will attend,...., 4 = definitely won't) | −0.12 (.13) | 1.11*** (.20) | 0.21 (.11) | 1.22*** (.20) | −0.89*** (.18) |
| Strength of votech attending plans—12th-grade (1 = definitely will attend,...., 4 = definitely won't) | −0.03 (.13) | −0.46** (.16) | 0.05 (.11) | −0.43** (.16) | 0.51*** (.13) |
| Have plans for the fall of 1969—12th-grade (1 = have plans, 0 = otherwise) | −0.12 (.35) | 0.13 (.51) | −0.05 (.30) | 0.25 (.52) | −0.18 (.45) |

*(Continued)*

| Independent variable | Military vs. work $b$ SE(b) | College vs. work $b$ SE(b) | Other vs. work $b$ SE(b) | College vs. military $b$ SE(b) | Other vs. military $b$ SE(b) | Other vs. college $b$ SE(b) |
|---|---|---|---|---|---|---|
| *Other* | | | | | | |
| Age | −0.004 | −0.35 | −0.31 | −0.34 | −0.30 | 0.05 |
| | (.32) | (.35) | (.28) | (.34) | (.27) | (.28) |
| Intellectual ability | 0.09* | 0.15** | 0.02 | −0.06 | −0.07 | −0.13*** |
| | (.05) | (.05) | (.04) | (.05) | (.04) | (.04) |
| Worked during 12th-grade | 0.001 | −0.29* | 0.04 | −0.29* | 0.05 | 0.33*** |
| (1 = no work, 2 = < 10 hours/week, 3 = >10 hours/week) | (.14) | (.14) | (.11) | (.14) | (.11) | (.10) |
| Intercept | −2.04 | 0.33 | .56 | .71 | 6.61 | 4.99 |
| | (6.96) | (6.48) | (5.26) | (6.32) | (5.01) | (5.03) |

Model Chi-Square = 784, df =75, N = 1,082

*p ≤ .05. **p ≤ .01. ***p ≤ .001.

*Note:* The *b* coefficients are maximum likelihood estimates. Their sign indicates whether the first context listed was more likely to be entered given the particular predictor variable (indicated by a positive *b*), or whether the second context was more likely to be entered (indicated by a negative *b*). More specifically: a positive *b* coefficient indicates that the first context listed was more likely to be chosen over the alternative context if the listed characteristic was present or as it increased in value. A negative *b* coefficient indicates that the first context listed was less likely to be chosen over the alternative context if the listed characteristic was present or as it increased in value. Conversely, the alternative context was more likely to be chosen if the listed characteristic was absent or as it decreased in value. A variable measuring whether or not the respondent dropped out of high school was excluded from the model because it did not contain sufficient variation to allow estimation to proceed.

# Appendix B. Variance-Covariance Matrix for Full Structural Equation Model of Social Contexts' Effects on Self-Esteem

| Variables | 1 | 2 | 3 | 4 | 5 | 6 | 7 | 8 | 9 | 10 |
|---|---|---|---|---|---|---|---|---|---|---|
| 1. I'm of worth—Wave 3 | .660 | | | | | | | | | |
| 2. I have good qualities—Wave 3 | .315 | .549 | | | | | | | | |
| 3. I do things well—Wave 3 | .238 | .245 | .526 | | | | | | | |
| 4. I have a positive self—Wave 3 | .258 | .232 | .238 | .659 | | | | | | |
| 5. I'm useful—Wave 3 | .163 | .205 | .159 | .174 | .499 | | | | | |
| 6. I do a job well—Wave 3 | .144 | .158 | .192 | .207 | .157 | .555 | | | | |
| 7. I'm of worth—Wave 5 | .163 | .127 | .122 | .108 | .077 | .093 | .472 | | | |
| 8. I have good qualities—Wave 5 | .145 | .165 | .135 | .141 | .088 | .097 | .295 | .478 | | |
| 9. I do things well—Wave 5 | .111 | .121 | .142 | .144 | .070 | .097 | .249 | .283 | .503 | |
| 10. I have a positive self—Wave 5 | .128 | .115 | .128 | .211 | .062 | .142 | .270 | .257 | .239 | .700 |
| 11. I'm useful—Wave 5 | .103 | .116 | .115 | .122 | .136 | .088 | .196 | .233 | .237 | .252 |
| 12. I do a job well—Wave 5 | .077 | .061 | .073 | .092 | .051 | .121 | .107 | .137 | .172 | .170 |
| 13. Selection bias term—Waves 1-5 | -.010 | -.009 | -.007 | -.007 | -.003 | .001 | -.004 | -.008 | -.004 | -.003 |
| 14. Intellectual ability—Wave 1 | .670 | .486 | .433 | .412 | -.189 | .052 | .523 | .599 | .469 | .392 |
| 15. Number of siblings—Wave 1 | -.090 | -.095 | -.038 | -.076 | -.012 | .039 | .020 | -.045 | .045 | .049 |
| 16. Work context | -.012 | -.019 | -.007 | -.004 | .011 | .008 | -.001 | -.013 | -.009 | -.006 |
| 17. Military context | -.020 | -.006 | -.010 | -.003 | -.006 | -.004 | -.012 | -.019 | -.001 | -.006 |
| 18. College context | .065 | .061 | .028 | .042 | .010 | .002 | .047 | .057 | .037 | .024 |
| 19. Time-in-context | .051 | .061 | -.005 | .072 | .006 | .045 | .108 | .075 | .068 | .052 |
| 20. Socioeconomic level—Wave 1 | .090 | .073 | .046 | .040 | -.002 | -.018 | .043 | .073 | .041 | .034 |
| 21. First born—Wave 1 | -.008 | -.005 | .003 | -.004 | -.002 | -.001 | .004 | -.012 | -.005 | .011 |
| 22. Reared on a farm—Wave 1 | -.016 | -.011 | -.003 | -.006 | .017 | -.006 | -.004 | -.006 | -.001 | -.005 |
| 23. Folks want S to work after H.S.—Wave 1 | -.008 | -.023 | -.013 | -.012 | -.001 | .006 | -.013 | -.009 | -.020 | -.022 |
| 24. Folks happy if S served in military after H.S.—Wave 1 | .040 | -.073 | -.081 | .075 | .119 | .004 | -.095 | -.017 | -.011 | -.050 |
| 25. GPA in 10th grade—Wave 1 | 1.088 | .958 | .867 | .789 | .280 | .327 | .618 | .770 | .645 | .554 |
| 26. College H.S track—Wave 3 | .030 | .032 | .019 | .021 | .014 | .009 | .022 | .033 | .021 | .009 |
| 27. Failed grade before 9th—Wave 1 | -.029 | -.015 | -.012 | -.014 | .006 | .013 | -.007 | -.018 | -.004 | -.007 |
| 28. Hawkish war attitudes—Wave 3 | .001 | .008 | .001 | -.021 | -.052 | -.023 | -.004 | .012 | .007 | -.012 |
| 29. S aspires to college—Wave 3 | .186 | .180 | .109 | .163 | .023 | .039 | .118 | .153 | .089 | .108 |
| 30. S aspires to tech school—Wave 3 | -.121 | -.076 | -.084 | -.054 | .017 | -.028 | -.090 | -.075 | -.045 | -.062 |
| 31. Hours worked in 12th grade—Wave 3 | -.038 | -.054 | -.034 | -.031 | .003 | .053 | .005 | -.023 | .014 | .009 |

| Variables | 11 | 12 | 13 | 14 | 15 | 16 | 17 | 18 | 19 | 20 |
|---|---|---|---|---|---|---|---|---|---|---|
| 11. I'm useful—Wave 5 | .540 | | | | | | | | | |
| 12. I do a job well—Wave 5 | .191 | .400 | | | | | | | | |
| 13. Selection bias term—Waves 1–5 | .000 | −.001 | .012 | | | | | | | |
| 14. Intellectual ability—Wave 1 | .144 | .081 | −.145 | 13.503 | | | | | | |
| 15. Number of siblings—Wave 1 | .146 | .042 | .049 | −1.521 | 3.420 | | | | | |
| 16. Work context | .003 | −.003 | .006 | −.211 | .066 | .109 | | | | |
| 17. Military context | .002 | .012 | .010 | −.171 | .113 | −.020 | .135 | | | |
| 18. College context | .025 | .009 | −.019 | .747 | −.223 | −.042 | −.053 | .224 | | |
| 19. Time-in-context | .071 | .066 | −.002 | .685 | −.035 | .146 | .152 | .247 | 1.788 | |
| 20. Socioeconomic level—Wave 1 | .017 | .005 | −.047 | 1.163 | −.373 | −.044 | −.053 | .125 | .025 | .577 |
| 21. First born—Wave 1 | .006 | −.007 | .006 | −.197 | .366 | .014 | .006 | −.030 | −.011 | −.050 |
| 22. Reared on a farm—Wave 1 | .013 | −.002 | .000 | −.153 | .074 | .005 | −.004 | −.006 | −.003 | −.042 |
| 23. Folks want S to work after H.S.—Wave 1 | −.010 | .001 | .004 | −.180 | .018 | .008 | .001 | −.019 | −.016 | −.030 |
| 24. Folks happy if S served in military after H.S.—Wave 1 | −.030 | .005 | .016 | −.835 | .473 | .073 | .042 | −.162 | −.018 | −.138 |
| 25. GPA in 10th grade—Wave 1 | .102 | .256 | −.397 | 14.070 | −2.319 | −.368 | −.405 | 1.494 | 1.194 | 1.286 |
| 26. College H.S track—Wave 3 | .004 | −.006 | −.009 | .290 | −.100 | −.012 | −.021 | .058 | .039 | .060 |
| 27. Failed grade before 9th—Wave 1 | .011 | .005 | .025 | −.415 | .101 | .007 | .034 | −.041 | .013 | −.068 |
| 28. Hawkish war attitudes—Wave 3 | −.020 | −.009 | −.006 | .346 | −.088 | −.005 | −.016 | .020 | .003 | .060 |
| 29. S aspires to college—Wave 3 | .049 | −.002 | −.045 | 1.881 | −.390 | −.089 | −.100 | .257 | .086 | .359 |
| 30. S aspires to tech school—Wave 3 | .005 | .010 | .019 | −1.119 | .262 | .036 | .053 | −.175 | −.142 | −.171 |
| 31. Hours worked in 12th grade—Wave 3 | −.005 | .009 | .011 | −.237 | .195 | .019 | .020 | −.070 | −.034 | −.080 |

| Variables | 21 | 22 | 23 | 24 | 25 | 26 | 27 | 28 | 29 | 30 |
|---|---|---|---|---|---|---|---|---|---|---|
| 21. First born—Wave 1 | .184 | | | | | | | | | |
| 22. Reared on a farm—Wave 1 | .011 | .099 | | | | | | | | |
| 23. Folks want S to work after high school—Wave 1 | −.001 | .000 | .088 | | | | | | | |
| 24. Folks happy if S served in military after H.S.—Wave 1 | .061 | −.016 | .034 | 9.521 | | | | | | |
| 25. GPA in 10th grade—Wave 1 | −.354 | −.017 | −.382 | −1.609 | 49.399 | | | | | |
| 26. College H.S track—Wave 3 | −.017 | −.011 | −.004 | −.009 | .669 | .209 | | | | |
| 27. Failed grade before 9th—Wave 1 | .009 | .002 | .010 | .013 | −.746 | −.021 | .125 | | | |
| 28. Hawkish war attitudes—Wave 3 | −.013 | −.016 | .000 | −.132 | .259 | .021 | −.022 | .326 | | |
| 29. S aspires to college—Wave 3 | −.063 | −.043 | −.054 | −.233 | 3.457 | .132 | −.107 | .049 | 1.225 | |
| 30. S aspires to tech school—Wave 3 | .026 | .032 | .041 | .216 | −1.974 | −.067 | .037 | −.051 | −.410 | .921 |
| 31. Hours worked in 12th grade—Wave 3 | .009 | −.006 | .016 | .086 | −.708 | −.002 | .024 | −.044 | −.106 | .070 |

| Variable | 31 |
|---|---|
| 31. Hours worked in 12th grade—Wave 3 | .874 |

# Appendix C. Phi Matrix of Correlations among the Exogenous Variables for Full Structural Equation Model of Social Contexts' Effects on Self-Esteem

| Variables | 13 | 14 | 15 | 16 | 17 | 18 | 19 | 20 | 21 | 22 |
|---|---|---|---|---|---|---|---|---|---|---|
| 13. Selection bias term—Waves 1–5 | 1.000 | | | | | | | | | |
| 14. Intellectual ability—Wave 1 | -.356 | 1.000 | | | | | | | | |
| 15. Number of siblings—Wave 1 | .240 | -.224 | 1.000 | | | | | | | |
| 16. Work context | .162 | -.174 | .108 | 1.000 | | | | | | |
| 17. Military context | .250 | -.127 | .167 | -.165 | 1.000 | | | | | |
| 18. College context | -.358 | .429 | -.254 | -.270 | -.302 | 1.000 | | | | |
| 19. Time-in-context | -.014 | .139 | -.014 | .331 | .309 | .390 | 1.000 | | | |
| 20. Socioeconomic level—Wave 1 | -.561 | .417 | -.266 | -.174 | -.192 | .349 | .025 | 1.000 | | |
| 21. First born—Wave 1 | .126 | -.125 | .462 | .097 | .038 | -.148 | -.020 | -.154 | 1.000 | |
| 22. Reared on a farm—Wave 1 | -.007 | -.133 | .127 | .052 | -.035 | -.043 | -.006 | -.176 | .079 | 1.000 |
| 23. Folks want S to work after high school—Wave 1 | .126 | -.166 | .032 | .082 | .008 | -.137 | -.041 | -.133 | -.006 | .002 |
| 24. Folks happy if S served in military after high school—Wave 1 | .048 | -.074 | .083 | .072 | .037 | -.111 | -.004 | -.059 | .046 | -.017 |
| 25. GPA in 10th grade—Wave 1 | -.509 | .545 | -.178 | -.158 | -.157 | .449 | .127 | .241 | -.117 | -.008 |
| 26. College H.S track—Wave 3 | -.177 | .172 | -.118 | -.082 | -.125 | .267 | .063 | .172 | -.084 | -.076 |
| 27. Failed grade before 9th—Wave 1 | .633 | -.320 | .154 | .056 | .265 | -.246 | .027 | -.252 | .063 | .019 |
| 28. Hawkish war attitudes—Wave 3 | -.097 | .165 | -.084 | -.024 | -.074 | .072 | .004 | .138 | -.054 | -.088 |
| 29. S aspires to college—Wave 3 | -.364 | .463 | -.190 | -.242 | -.247 | .491 | .058 | .427 | -.132 | -.123 |
| 30. S aspires to tech school—Wave 3 | .178 | -.317 | .148 | .113 | .150 | -.386 | -.111 | -.235 | .062 | .107 |
| 31. Hours worked in 12th grade—Wave 3 | .108 | -.069 | .113 | .062 | .060 | -.158 | -.027 | -.113 | .024 | -.020 |

| Variable | 23 | 24 | 25 | 26 | 27 | 28 | 29 | 30 | 31 |
|---|---|---|---|---|---|---|---|---|---|
| 23. Folks want S to work after high school—Wave 1 | 1.000 | | | | | | | | |
| 24. Folks happy if S served in military after H.S.—Wave 1 | .037 | 1.000 | | | | | | | |
| 25. GPA in 10th grade—Wave 1 | -.183 | -.074 | 1.000 | | | | | | |
| 26. College H.S track—Wave 3 | -.030 | -.006 | .208 | 1.000 | | | | | |
| 27. Failed grade before 9th—Wave 1 | .092 | .012 | -.300 | -.128 | 1.000 | | | | |
| 28. Hawkish war attitudes—Wave 3 | -.002 | -.075 | .065 | .079 | -.111 | 1.000 | | | |
| 29. S aspires to college—Wave 3 | -.165 | -.068 | .444 | .260 | -.273 | .078 | 1.000 | | |
| 30. S aspires to tech school—Wave 3 | .143 | .073 | -.293 | -.153 | .110 | -.094 | -.386 | 1.000 | |
| 31. Hours worked in 12th grade—Wave 3 | .057 | .030 | -.108 | -.006 | .072 | -.083 | -.103 | .078 | 1.000 |

# References

Aldrich, J. H., & Nelson, F. D. (1984). *Linear probability, logit, and probit models.* Beverly Hills, CA: Sage.

Aldrich, J. H., & Nelson, F. D. (1984). *Logit and probit models for multivariate analysis with qualitative dependent variables.* In W. D. Berry, & M. S. Lewis-Beck (Eds.), *New tools for social scientists: Advances and applications in research methods* (pp. 115–155). Beverly Hills, CA: Sage.

Alexander, T. (1982). The life course issues. *Academic Psychology Bulletin,* 515–526.

Allport, G. W. (1961). *Pattern and growth in personality.* New York: Holt, Rinehart and Winston.

Alwin, D. F., Cohen, R. L., & Newcomb, T. M. (1991). *Political attitudes over the life span: The Bennington women after fifty years.* Madison, WI: University of Wisconsin Press.

Alwin, D. F., & McCammon. (2003). Generation, cohorts, and social change. In J. T. Mortimer, & M. J. Shanahan (Eds.), *Handbook of the Life Course* (pp. 23–49). New York: Kluwer Academic/Plenum Publishers.

Anderson, A. B., Basilevsky, A., & Hum, D. P. J. (1983). Missing data: A review of the literature. In P. H. Rossi, J. D. Wright, & A. B. Anderson (Eds.), *Handbook of survey research* ( pp. 231–287). New York : Academic Press.

Andreski, S. (1968). *Military organization and society.* Berkeley: University of California.

Astin, A. W. (1977). *Four critical years: Effects of college on beliefs, attitudes, and knowledge.* San Francisco: Jossey-Bass.

Bachman, J. G. (1975). *Youth in transition.* Ann Arbor, MI: Survey Research Center, Institute for Social Research, The University of Michigan.

Bachman, J. G. (1970). *Youth in transition: The impact of family background and intelligence on tenth-grade boys* (Vol 2. ed.,). Ann Arbor: Institute for Social Research.

Bachman, J. G., Kahn, R. L., Mednick, M. J., Davidson, T. N., & Johnston, L. D. (1967). *Youth in transition: Blueprint for a longitudinal study of adolescent boys.* Ann Arbor: Institute for Social Research.

Bachman, J. G., O'Malley, P. M., & Johnston, J. (1978). *Youth in transition: Adolescence to adulthood– change and stability in the lives of young men* Vol. 6 ed.. Ann Arbor, MI: Institute for Social Research.

Baltes, P. B. (1979). Life-span developmental psychology: Some converging observations on history and theory. In P. B. Baltes, & O. G. Jr. Brim (Eds.), *Handbook of the psychology of aging* (Vol. 2pp. 21–38). New York: Academic Press.

173

Bandura, A. (1982). The self and mechanisms of aging. In T. Suls (Ed.), *Psychological perspectives on the self* (pp. 3–39). Hillsdale, NJ: Lawrence Erlbaum.

Bandura, A. (1997). *Self-efficacy: The exercise of control.* New York: W.H. Freeman.

Bandura, A. (1977). Self-efficacy: Toward a unifying theory of behavioral change. *Psychological Review, 84,* 191–215.

Bandura, A. (1981). Self-referent thought: A developmental analysis of 'self efficacy'. In I. N. Flavell, & L. Ross (Eds.), *Social cognitive development: Frontiers and possible futures* (pp. 200–239). New York: Cambridge University Press.

Beck, A. T. (1967). *Depression: Causes and treatment.* Philadelphia: University of Pennsylvania Press.

Bengston, V. L., Reedy, M. N., & Gordon, C. (1985). *Aging and self-conceptions: Personality processes and social contexts.* In J. E. Birren, & K. W. Schaie (Eds.), *Handbook of psychology of aging* (2nd ed., pp. 544–593). New York: Van Nostrand Reinhold.

Bengtson, V. L., Biblarz, T. J., & Roberts, R. E. L. (2002). *How families still matter: A longitudinal study of youth in two generations.* Cambridge, UK: Cambridge University Press.

Bentler, P. M. (1992). *EQS: Structural equations program manual.* LA, CA: BMDP Statistical Software.

Bentler, P. M., & Bonett, D. G. (1980). Significance tests and goodness of fit in the analysis of covariance structures. *Psychological Bulletin, 88,* 588–606.

Berk, R. A. (1983). An introduction to sample selection bias in sociological data. *American Sociological Review, 48,* 191–215.

Berry, W. D., & Feldman, S. (1985). Multiple regression in practice. Beverly Hills, CA: Sage.

Blake, J. (1989). *Family size and achievement.* Berkeley, CA: University of California Press.

Blake, J. (1981). Family size and the quality of children. *Demography, 18,* 421–442.

Blau, P. M., Gustad, J. W., Jessor, R., Parnes, H. S., & Wilcock, R. C. (1956). Occupational choice: A conceptual framework. *Industrial and Labor Relations Review, 9,* 531–543.

Bollen, K. A. (1989). *Structural equations with latent variables.* New York: John Wiley and Sons.

Bombard, J. (1981). The Screaming eagles. In A. Santoli (Ed.), *Everything we had: An oral history of the Vietnam War by thirty-three American soldiers who fought it* (pp. 106–109). New York: Random House.

Booth, A., & Johnson, D. R. (1985). Tracking respondents in a telephone interview panel selected by random digit dialing. *Sociological Methods and Research, 14,* 53–64.

Borman, K. M. (1988). The process of becoming a worker. In K. M. Borman, & J. T. Mortimer (Eds.), *Work experience and psychological development through the life course* (pp. 51–75). Boulder, CO: Westview Press.

Borman, K. M., & Hopkins, M. C. (1987). Leaving school for work. In R. G. Corwin (Ed.), *Research in sociology of education and socialization, Vol. 7* (pp. 131–179). Greenwich, CT: JAI.

Borooah, V. K. (2002). *Logit and probit: Ordered and multinomial models* Vol. Sage University Paper Series on Quantitative Applications in the Social Sciences, 07-138. Thousand Oaks, CA: Sage.

Borus, M. E., & Carpenter, S. A. (1984). Choics in education. In M. E. Borus (Ed.), *Youth and the labor market* (pp. 81–110). Kalamazoo, MI: W.E. Upjohn Institute for Employment Research.

Boulanger, G. (1981). Who goes to war? In A. Egendorf, C. Kadushin, R. S. Laufer, G. Rothbart, & L. Sloan (Eds.), *Long term stress reactions: Some causes, consequences, and naturally occuring support systems* (Vol. IV pp. 495–515). Washington, D.C.: U.S. Government Printing Office.

Bourricaud, F. (1977). *The sociology of Talcott Parsons* Goldhammer, Arthur (trans., ed.). Chicago: University of Chicago.

Bowen, H. R. (1977). *Investment in learning: The individual and social value of American higher education.* San Francisco: Jossey-Bass.

Bowles, S., & Gintis, H. (1968). *Schooling in capitalist America.* New York: Basic Books.

Bradburn, E. M., Moen, P., & Dempster-Mcclain, D. (1995). Women's return to school following the transition to motherhood. *Social Forces, 73,* 1517–1551.

Breen, R. (1996). *Regression models: Censored, sample selected, or truncated data* Vol. Sage University Paper Series on Quantitative Applications in the Social Sciences, 07-111. Thousand Oaks, CA: Sage.

Brim, O. G. Jr. (1966). Socialization through the life course. In O. G. Jr. Brim, & S. Wheeler (Eds.), *Socialization after high school* (pp. 3–49). New York: John Wiley & Sons.

Brim, O. G. Jr., & Kagan, J. (1980). Constancy and Change in human Development. Cambridge: Harvard University Press.

Bronfenbrenner, U. (1979). *The ecology of human development*. Cambridge, MA: Harvard University Press.

Buehler, C. (1933). Der Menschliche Lebenslauf als Psychologisches Problem. Leipzig, Germany: Hirzel.

Burns, R. B. (1979). *The self concept in theory, measurement, development, and behaviour*. London: Longman.

Bush, D., & Simmons, R. G. (1990). Socialization processes over the life course. In M. Rosenberg, & R. H. Turner (Eds.), *Social psychology: Sociological perspectives* (revised ed., pp. 133–164). New Brunswick: Transaction.

Carlsson, G., & Karlsson, K. (1970). Age, cohorts and the generation of generations. *American Sociologica; Review, 35*, 710–718.

Carmines, E. G., & McIver, J. P. (1981). Analyzing models with unobserved variables: Analysis of covariance structures. In G. W. Bohrnstedt, & E. F. Borgattta (Eds.), *Social measurement: Current issues* (pp. 65–115). Beverly Hills: Sage.

Carmines, E. G., & Zeller, R. A. (1979). *Reliability and validity assessment*. Beverly Hill: Sage.

Caspi, A. (2000). The child is father of the man: Personality continuities from childhood to adulthood. *Journal of Personality & Social Psychology, 78*, 158–172.

Cherlin, A., & Reeder, L. G. (1975). The dimensions of psychological well-being: A critical review. *Sociological Methods and Research, 4*, 189–214.

Clausen, J. A. (1991). Adolescent competence and the shaping of the life course. *American Journal of Sociology, 96*, 805–842.

Clausen, J. A. (1993). *American lives: Looking back at the children of the Great Depression*. New York: Free Press.

Clausen, J. A. (1986). *The life course: A sociological perspective*. Engelwood Cliff, NJ: Prentice-Hall.

Clausen, J. A., & Clausen, S. R. (1973). The effect of family size on parents and children. In J. T. Fawcett (Ed.), *Psychological perspectives on population* (pp. 185–208). New York: Basic Books.

Cobb, S., Brooks, G. H., Kasl, S. V., & Connelly, W. E. (1966). The health of people changing jobs: A description of a longitudinal study. *American Journal of Public Health, 56*, 1476–1481.

Cohen, J., & Cohen, P. (1983). *Applied multiple regression/correlation analysis for the behavioral sciences* 2nd ed.. Hillsdale, NJ: Lawrence, Erlbaum.

Conger, R. D., & Elder, G. H. Jr. (1994). Families in troubled times: Adapting to change in rural America. Hawthorne, NY: Aldine de Gruyter.

Cooley, C. H. (1922). *Human nature and the social order*. New York: Charles Scribner's Sons.

Cooley, C. H. (1909). Social organization: A study of the larger mind. New York: Scribner's Sons.

Coopersmith, S. (1967). *The antecedents of self-esteem*. San Francisco: Freeman.

Corsaro, W. A., & Eder, D. (1995). Development and socialization in children and adults. In K. S. Cook, G. A. Fine, & J. S. House (Eds.), *Sociological perspectives on social psychology* (pp. 421–451). Boston: Allyn & Bacon.

Crandall, R. (1973). The measurement of self-esteem and related constructs. In J. P. Robinson, & P. R. Shaver (Eds.), *Measures of social psychological attitudes* (pp. 45–167). Ann Arbor: Institute for Social Research.

Dannefer, D. (1984). Adult development and social theory: A paradigmatic reappraisal. *American Sociological Review, 49*, 100–116.

Davis, J. A. (1968). The campus as a frog pond: An application of the theory of relative deprivation to career decisions of college men. *American Journal of Sociology, 72,* 17–31.

Department of Defense. (1972). *Computer study of casualties in Vietnam.* Washington, D.C.: U.S. Government Printing Office.

Dreeben, R. (1968). *On what is learned in school.* Reading, MA: Addison-Wesley.

Dube, S. R., Felitti, V. J., Dong, M., Giles, W. H., & Anda, R. F. (2003). The impact of adverse childhood experiences on health problems: Evidence from four birth cohorts dating back to 1900. *Preventive Medicine, 37,* 268–277.

Economic History Services. (2004). Web site: URL http://www.eh.net/ehresources/howmuch/inflationq.php

Eder, D., & Nenga, S. K. ( 2003). Socilization in adolescence. In J. Delamater (Ed.), *Handbook of social psychology* (pp. 157–182). New York: Kluwer Academic/Plenum Publishers.

Edwards, V. J., Holden, G. W., Felitti, V. J., & Anda, R. F. (2003). Relationship between multiple forms of childhood maltreatment and adult mental health in community respondents: Results from the Adverse Childhood Experiences study. *American Journal of Psychiatry, 160,* 1453–1460.

Egendorf, A., Kadushin, C., Laufer, R. S., Rothbart, G., & Sloan, L. (1981). *Legacies of Vietnam: Comparative adjustment of veterans and their peers.* New York: US Government Printing Office.

Eisenhart, R. W. (1975). You can't hack it little girl: A discussion of the covert psychological agenda of modern combat training. *Journal of Social Issues, 31,* 13–23.

Elder, G. H. Jr. (1999). *Children of the great depression: Social change in life experience* 25th Anniversary ed.. Boulder, CO: Westview Press.

Elder, G. H. Jr. (1998). The life course as developmental theory. *Child Development, 69,* 1–12.

Elder, G. H. Jr. (1986). Military times and turning points in men's lives. *Developmental Psychology, 22,* 233–245.

Elder, G. H. Jr. (1987). War mobilization and the life course: A cohort of World War II veterans. *Sociological Forum, 2,* 449–472.

Elder, G. H. J., & Rockwell, R. C. (1979). Economic depression and postwar opportunity in men's lives: A study of life patterns and health. *Research in Community and Mental Health, 1,* 249–303.

Elder, G. H. J., & Rockwell, R. C. (1977). *The life course and human development: An ecological perspective.* Omaha, NE: The Boys Town Center for the Study of Youth Development.

Elder, G. H. Jr., & Clipp, E. C. (1989). Combat experience and emotional health: Impairment and resilience in later life. *Journal of Personality, 57,* 311–341.

Elder, G. H. Jr., Johnson, M. K., & Crosnoe, R. (2003). The emergence and development of life course theory. In J. T. Mortimer, & M. Shanahan (Eds.), *Handbook of the life course* (pp. 3–19). New York: Kluwer Academic/Plenum Publishers.

Entwisle, D. R., Alexander, K. L., & Olson, L. S. (2000). Early work histories of urban youth. *American Sociological Review, 65,* 279–297.

Erikson, E. H. (1963). *Childhood and society* 2nd ed.. New York: W. W. Norton and Co.

Erikson, E. H. (1968). *Identity: youth and crisis.* New York: W. W. Norton.

Erikson, E. H. (1982). *The Life Cycle Completed.* New York: W. W. Norton.

Ernst, C., & Angst, J. (1983). *Birth order: Its influence on personality.* Berlin: Springer-Verlag.

Faris, J. H. (1976). Socialization into the all-volunteer force. In N. L. Goldman, & D. R. Segal (Eds.), *The social psychology of military service* (pp. 13–24). Beverly Hills: Sage.

Feldman, K. A., & Newcomb, T. M. (1973). The impact of college on students. Vol. 1 & 2. San Francisco: Jossey-Bass.

Figley, C. A., & Leventman, S. (ed). (1980). *Strangers at home: Vietnam veterans since the war.* New York: Praeger.

Finch, M. D., & Mortimer, J. T. (1985). Adolescent work hours and the process of achievement. In A. C. Kerckhoff (Ed.), *Research in sociology of education and socialization* (Vol. 5). Greenwich, CT: JAI.

Fleming, J. S., & Courtney, B. E. (1984). The Dimensionality of self-esteem: II. Hierarchical facet model for revised measurement scales. *Journal of Personality and Social Psychology, 46,* 404–421.

Fligstein, N. D. (1980). Who served in the military, 1940–1973? *Armed Forces and Society, 6,* 297–312.

Foner, A. (1978). Ascribed and achieved bases of stratification. *Annual Review of Sociology, 5,* 219–242.

Foucault, M. (2003). The essential Foucault: Selections from the essential works of Foucault, 1954–1984. In P. Rabinow, & N. Rose (Eds.). New York: New Press.

Freedman, M. B. (1979). What happens after college: Studies of alumni. In N. Sanford, & J. Axelrod (Eds.), *College and character* (pp. 129–135). Berkeley: Montaigne.

Funk, R. B., & Willits, F. K. (1987). College attendance and attitude change: A panel study, 1970–1981. *Sociology of Education, 60,* 224–231.

Gabriel, R. A., & Savage, P. L. (1978). Crisis in command: Mismanagememt in the army. New York: Hill and Wang.

Gecas, V. (1981). Contexts of socialization. In M. Rosenberg, & R. H. Turner (Eds.), *Social psychology: Sociological perspectives* (pp. 165–199). New York: Basic Books.

Gecas, V. (1979). The Influence of social class on socialization. In W. R. Burr, R. Hill, F. I. Nye, & I. L. Reiss (Eds.), *Contemporary theories about the family: Research-based theories* (Vol. 1 ed., pp. 365–404). New York: Free Press.

Gecas, V. (2003). Self-agency and the life course. In J. T. Mortimer, & M. J. Shanahan (Eds.), *Handbook of the life course* (pp. 369–388). New York: Kluwer Academic/Plenum Publishers.

Gecas, V. (1982). The Self-concept. *Annual Review of Sociology, 8,* 1–33.

Gecas, V. (1989). The Social psychology of self-efficacy. *Annual Review of Sociology, 15,* 291–316.

Gecas, V. (2000). Socialization. In E. F. Borgatta (Ed.), *Encyclodedia of sociology* (pp. 2855–2864). Detroit: Macmillan.

Gecas, V., & Schwalbe, M. L. (1983). Beyond the looking-glass self: Social structure and eEfficacy-bBased sSelf-eEsteem. *Social Psychology Quarterly, 46,* 77–88.

George, L. K. (1993). Sociological perspectives on life transitions. *Annual Review of Sociology, 19,* 353–373.

Gergen, K. J. (1971). *The concept of self.* New York: Holt, Rinehart, and Winston.

Gergen, K. J. (1980). The emerging crisis in life-span developmental theory. In P. B. Baltes, & O. G. Jr. Brim (Eds.), *Constancy and change in human development* (pp. 201–231). Cambridge: Harvard University Press.

Glenn, N. D. (2003). Distingushiing age, period, and cohort effects. In J. T. Mortimer, & M. J. Shanahan (Eds.), *Handbook of the life course* (pp. 465–476). New York: Kluwer Academic/Plenum Publishers.

Glenn, N. D. (1980). Values, attitudes, and beliefs. In O. G. Jr. Brim, & K. Kagan (Eds.), *Constancy and change in human development* (pp. 596–640). Cambridge: Harvard University Press.

Goffman, E. (1961). *Asylums: Essays on the social situation of mental patients and other inmates.* Garden City, NY: Doubleday Anchor.

Goldsmith, R. E. (1986). Dimensionality of the Rosenberg self-esteem scale. *Journal of Social Behavior and Personality, 1,* 253–264.

Gordon, C. (1972). Looking ahead: Self-conceptions, race, and family as determinants of adolescent orientation to achievement. Washington, DC: American Sociological Association.

Granovetter, M. (1983). The strength of weak ties: A network theory revisited. *Sociological Theory, 1,* 201–233.

Granovetter, M. S. (1974). *Getting a job: A study of contacts and careers.* Cambridge, MA: Harvard University Press.

Granovetter, M. S. (1973). The strength of weak ties. *American Journal of Sociology, 78,* 1360–1380.

Greene, W. H. (2002). *LIMDEP 8.0.* New York: Econometric Software, Inc.

Greenwald, A. G. (1980). The totalitarian ego: Fabrication and revision of personal history. *American Psychologist, 7*, 603–618.

Gunderson, E. E. K. (1976). Health and adjustment of men at sea. In N. L. Goldman, & D. R. Segal (Eds.), *The social psychology of military service* (pp. 67–88). Beverly Hills: Sage.

Han, S.-K., & Moen, P. (1999). Work and family over time: A life course approach. *Annals of the American Academy of Political & Social Science, 56*, 98–110.

Hareven, T. K. (1981). Historical changes in the timing of family transitions: Their impact on generation relations. In R. W. Fogel, Hatfield Elaine, S. B. Kiesler, & E. Shanas (Eds.), *Aging: Stability and change in the family* (pp. 41–62). New York: Academic Press.

Harris, M. B., Benson, S. M., & Hall, C. L. (1975). The effects of confession on altruism. *Journal of Social Psychology, 96*, 187–192.

Harter, S. (1985). Competence as a dimension of self-evaluation: Toward a comprehensive model of self-worth. In R. L. Leahy (Ed.), *The development of the self* (pp. 55–121). Orlando, FL: Academic Press.

Heer, D. M. (1985). Effects of sibling number on child outcome. *Annual Review of Sociology, 11*, 27–47.

Heider, F. (1958). *The psychology of interpersonal relations*. New York: Wiley.

Heise, D. R. (1970). Causal inference from panal data. In E. F. Borgatta, & G. W. Bohrnstedt (Eds.), *Sociological methodology*. San Francisco, CA: Jossey-Bass.

Heiss, J. (1990). Social roles. In M. Rosenberg, & R. H. Turner (Eds.), *Social psychology: Sociological perspectives* (revised ed., pp. 94–129). New Brunswick, NJ: Transaction.

Hendin, H., & Pollinger Haas, A. (1984). *Wounds of war: The psychological aftermath of combat in Vietnam*. New York: Basic Books.

Hogan, D. P. (1981). *Transitions and social change: The early lives of American men*. New York: Academic Press.

Holland, J. L. (1997). *Making vocational choices: A theory of vocational personalities and work environments* 3rd ed.. Odessa, FL: Psychological Assessment Resources.

Horn, J. L., & Donaldson, G. (1980). Cognitive development in adulthood. In O. G. Jr. Brim, & J. Kagan (Eds.), *Constancy and change in human development* (pp. 62–84). Cambridge: Harvard University Press.

Horney, K. (1950). *Neurosis and human growth*. New York: Norton.

House, J. S. (1981). Social structure and personality. In M. Rosenberg, & R. H. Turner (Eds.), *Social psychology: Sociological perspectives* (pp. 525–561). New York: Basic Books.

Janowitz, M. (1964). *The professional soldier: A social and political portrait*. New York: Free Press.

Janowitz, M. (1971). *The professional soldier: A social and political portrait*. New York: Free Press.

Janowitz, M., & Little, R. (1965). *Sociology and the military establishment* revised ed.. New York: Russell Sage.

Johnston, J., & Bachman, J. G. (1972). *Youth in transition: Young men and military service* Vol. 5. Ann Arbor, MI: Institute for Social Research.

Joreskog, K., & Sorbom, D. (1989). *LISREL 7: A guide to the program and applications* 2nd ed.. Chicago: SPSS.

Kadushin, C., Goulanger, G., & Martin, J. (1981). Long term stress reactions: Some causes, consequences, and naturally occuring support systems. *Legacies of Vietnam* Vol. 5 ed.. Washington, D.C.: U.S. Government Printing Office.

Kanouse, D. E., Haggstrom, G. W., Blaschke, T. J., Kaham, J. P., Lisowski, W., & Morrison, P. A. (1980). *Effects of postsecondary experience on aspirations, attitudes, and self-conceptions*. Santa Monica, CA: RAND.

Kanter, R. M. (1977). *Men and women of the corporation*. New York: Basic Books.

Kaplan, H. B. (1975). *Self-attitudes and deviant behavior*. Pacific Palisades, CA: Goodyear.

Kaplan, H. B. (1971). Social class and self-derogation: A conditional relationship. *Sociometry, 34*, 41–65.

Kaplan, H. B., & Pokorny, A. D. (1969). Self-derogation and psychosocial adjustment. *Journal of Nervous and Mental Disease, 149*, 421–434.

Kaysen, C. (1969). *The higher learning, the universities, and the public*. Princeton, NJ: Princeton University Press.

Kelley, H. H. (1971). *Attribution in social interaction*. Morristown, NJ: General Learning.

Kemper, T. D. (1987). How many emotions are there? Wedding the social and the autonomic components. *American Journal of Sociology, 93*, 263–289.

Kennedy, P. (1998). *A guide to econometrics* 4th ed.. Cambridge, MA: MIT Press.

Kerckhoff, A. C. (1974). *Ambition and attainment: A study of four samples of American boys*. Washington D.C.: American Sociological Association.

Kerckhoff, A. C. (1993). *Diverging pathways: social structure and career deflections*. New York: Cambridge University Press.

Kohn, M. L. (1977). *Class and conformity: A study in values* 2nd ed.. Chicago: University of Chicago Press.

Kohn, M. L. (1969). *Class and conformity: A study of values*. Homewood, IL: Dorsey.

Kohn, M. L., & Schooler, C. (1973). Occupational experience and psychological functioning: An assessment of reciprocal effects. *American Journal of Sociology, 38*, 97–118.

Kohn, M. L., & Schooler, C. (1983). *Work and personality: An inquiry into the impact of social stratification*. Norwood, NJ: Ablex.

Kohn, M. L., Zaborowski, W., Janicka, K., Khmelko, V., Mach, B. W., Paniotto, V. et al. (2002). Structural location and personality during the transformation of Poland and Ukraine. *Social Psychology Quarterly, 65*, 364–385.

Kohn, M. L., Zaborowski, W., Janicka, K., Mach, B. W., Khmelko, V., Slomczynski, K. M. et al. (2000). Complexity of activities and personality under conditions of radical social change: A comparative analysis of Poland and Ukraine. *Social Psychology Quarterly, 63*, 187–207 .

Krepinevich, A. (1986). The army and Vietnam. Baltimore: Johns Hopkins University Press.

Ladd, E. C. Jr., & Lipset, S. M. (1975). *The divided academy: Professors and politics*. New York: McGraw-Hill.

Laufer, R. S., Gallops, M. S., & Frey-Wouters, E. (1982). *War stress and trauma: The Vietnam veteran experience*. New York: Center for Policy Research.

Laufer, R. S., Yager, T., Frey-Wouters, E., & Dommellan, J. (Post-war trauma: Social and psychological problems of Vietnam veterans in the aftermath of the Vietnam War). (1981). *Legacies of Vietnam* Vol. 3 ed. Washington D.C.: U.S. Government Printing Office.

Lecky, P. (1951). Self-consistency: A theory of personality. Shoe String Press.

Levinson, D. J. (1978). *The season's of a man's life*. New York: Ballantine Books.

Levinson, D. J. (1996). *The seasons of a woman's life*. New York: Knopf.

Linton, R. (1936). *The study of man: An introduction*. New York: D. Appleton-Century.

Long, J. S. (1983). (Confirmatory factor analysis: A preface to LISREL). *Sage university paper series on quantitative applications in the social sciences, series no. 07-033*. Beverly Hills: Sage.

Lowman, J., Galinsky, D., & Gray-Little, B. (1980). *Predicting achievement: A ten-year followup on black and white Adolescents*. Chapel Hill, NC: The University of North Carolina at Chapel Hill, Institute for Research in Social Sciences.

Lutfey, K., & Mortimer, J. T. (2003). Development and socialization through the adult life course. In J. Delamater (Ed.), *Handbook of social psychology* (pp. 183–202). New York: Kluwer Academic/Plenum Publishers.

Maddala, G. S. (1983). *Limited-dependent and qualitative variables in econometrics*. Cambridge, Eng.: Cambridge University Press.

Mannheim, K. (1952). The problem of generation. In P. Kecskemeti (Ed.), *Essays on the sociology of knowledge* (pp. 276–320). London: Routledge and Kegan Paul.

Markus, H., & Wurf, E. (1987). The dynamic self-concept: A social psychological perspective. *Annual Review of Psychology, 38*, 299–337.

Martindale, M., & Poston, D. L. J. (1979). Variations in veteran/nonveteran earnings patterns among World War II, Korea, and Vietnam War cohorts. *Armed Forces and Society, 5*, 219–243.

Marx, K. (1964). In T. B. Bottomore (Ed.), *Selected works*. New York: McGraw-Hill.

Mayhew, B. H. (1980). Structuralism versus individualism: Part I, shadowboxing in the dark. *Social Forces, 59*, 335–375.

McConnell, T. M., Heist, P., & Axelrod, J. (1979). The diverse college student population. In N. Sanford, & J. Axelrod (Ed.), *College and character* (pp. 69–83). Berkeley: Montaigne.

Mead, G. H. (1934). *Mind, self, and society form the standpoint of a social behaviorist*. Chicago, IL: University of Chicago Press.

Merton, R. K. (1976). *Sociological ambivalence and other essays*. New York: Free Press.

Miller, J., & Garrison, H. H. (1982). Sex roles: The division of labor at home and in the workplace. *Annual Review of Sociology, 8*, 237–262.

Mirowsky, J., & Reynolds, J. R. (2000). Age, depression, and attrition in the national survey of families and households. *Sociological Methods & Research, 28*, 476–504.

Moen, P. (2003). Midcourse: Navigating Retirement and a New Life Stage. In J. T. Mortimer, & M. J. Shanahan (Eds.), *Handbook of the Life Course* (pp. 269–291). New York: Kluwer Academic/Plenum Publishers.

Moen, P., & Dempster-McClain, D. (1989). Social integration and longevity: An event eistory analysis of women's roles and resilience. *American Sociological Review, 54*, 635–647.

Mortimer, J. T. (1996). Social psychological aspects of achievement. In A. C. Kerckhoff (Ed.), *Generating social stratification: Toward a new generation of research* (pp. 17–36). Boulder, CO: Westview.

Mortimer, J. T. (2003). *Working and growing up in america*. Cambridge, MA: Harvard University Press.

Mortimer, J. T., Finch, M., Shanahan, M., & Ryu, S. (1992). Work experience, mental health, and behavioral adjustment in adolescence. *Journal of Research on Adolescence, 2*, 25–57.

Mortimer, J. T., & Finch, M. D. (1986). The effects of part-time work on adolescent self-concept and achievement. In K. Borman, & J. Reisman (Eds.), *Becoming a worker* (pp. 66–89). Norwood, NJ: Ablex.

Mortimer, J. T., Finch, M. D., & Kumka, D. (1982). Persistence and change in development: The multidimensional self-concept. *Life-Span Development and Behavior, 4*, 263–313.

Mortimer, J. T., & Lorence, J. (1979). Occupational experience and the self-concept: A longitudinal study. *Social Psychology Quarterly, 42*, 307–323.

Mortimer, J. T., Lorence, J., & Kumka, D. S. (1986). *Work, family, and personality: Transition to adulthood*. Norwood, NJ: Ablex.

Mortimer, J. T., & Shanahan, M. (2003). Handbook of the life course. (Eds.). New York: Kluwer Academic/Plenum Publishers.

Mortimer, J. T., & Simmons, R. G. (1978). Adult socialization. *Annual Review of Sociology, 4*, 421–454.

Moskos, C. C. Jr. (1970). *The American enlisted man: The rank and file in today's military*. New York: Russell Sage Foundation.

Moskos, C. C. Jr. (1976). The military. *Annual Review of Sociology, 2*, 55–77.

Murphy, G. (1947). *Personality: A biosocial approach to origins and structure*. New York: Harper.

Murray, S. O., Rankin, J. H., & Magill, D. W. (1981). Strong ties and job information. *Sociology of Work and Occupations, 8*, 119–135.

Newcomb, T. M. (1965). Attitude development as a function of reference groups: The Bennington study. In H. M. Prashansky, & B. Seidenberg (Eds.), *Basic studies of social psychology* (pp. 215–224). New York: Holt, Rinehart, and Winston.

Newcomb, T. M. (1943). *Personality and social change: Attitude formation in a student community.* New York: Holt, Rinehart, and Winston.

Newcomb, T. M. (1979). Student peer-group influence. In N. Sanford, & J. Axelrod (Eds.), *College and character* (pp. 141–145). Berkeley, CA: Montaigne.

O'Malley, P. M., & Bachman, J. G. (1983). Self-esteem: Change and stability between Ages 13 and 23. *Developmental Psychology, 19*, 257–268.

O'Rand, A. M. (1996). Context, selection and agency in the life course: Linking social structure and biography. In A. Weymann, & W. R. Heinz (Eds.), *Society and biography: Interrelationships between social structure, institutions, and the life course* (pp. 67–81). Weinheim: Deutscher Studien Verlag.

O'Rand, A. M. (1995). The cumulative satisfaction of the life course. In R. H. Binstock, & L. K. George (Eds.), *The handbook of aging and the social sciences* (4th ed., pp. 188–207). San Diego, CD: Academic Press.

O'Rand, A. M. (1996). The precious and the precocious: Understanding cumulative disadvantage and cumulative advantage over the life course. *The Gerontologist, 36*, 230–238.

Osipow, S. H. (1968). Theories of career development. New York: Appleton-Century-Crofts.

Owens, T. J. (1993). Accentuate the positive-and the negative: Rethinking the use of self-esteem, self-deprecation, and self-confidence. *Social Psychology Quarterly, 56*, 288–299.

Owens, T. J. (1992). The effect of post-high school social context on self-esteem. *Sociological Quarterly, 33*, 553–577.

Owens, T. J. (2003). Self and identity. In J. D. Delamater (Ed.), *Handbook of social psychology* (pp. 205–232). New York: Kluwer Academic/Plenum.

Owens, T. J. (ed). (2000). *Self and identity through the life course in cross-cultural perspective.* Stamford, Connecticut: JAI.

Owens, T. J. (1994). Two dimensions of self-esteem: Reciprocal effects of positive self-worth and self-deprecation on adolescent problems. *American Sociological Review, 59*, 391–407.

Owens, T. J. (1992). Where do we go from here? Post-high school choices of American men. *Youth and Society, 23*, 452–477.

Owens, T. J., & King, A. B. (2001). Measuring self-esteem: Race, ethnicity, and gender considered. In T. J. Owens, S. Stryker, & N. Goodman (Eds.), *Extending self-esteem theory and research: Sociological and psychological currents* (pp. 56–84). New York: Cambridge University.

Owens, T. J., Mortimer, J. T., & Finch, M. D. (1996). Self-determination as a source of self-esteem in adolescence. *Social Forces, 74*, 1377–1404.

Owens, T. J., & Serpe, R. (2003). The role of self-esteem in family identity salience and commitment among African-Americans, Latinos, and Whites. In P. J. Burke, T. J. Owens, R. Serpe, & P. A. Thoits (Eds.), *Advances in identity theory and research* (pp. 85–104). New York: Kluwer Academic/Plenum Publishers.

Panel on Youth of the President's Science Advisory Committee. (1974). Youth: Transition to adulthood. Chicago: University of Chicago Press.

Park, R. E., & Burgess, E. W. (1921). Introduction to the science of sociology. Chicago: University of Chicago Press.

Parsons, T. (1942). Age and sex in the social structure of the United States. *American Sociological Review, 7*, 604–616.

Pavalko, E. K. (1997). Beyond trajectories: Multiple concepts for analyzing long-term process. In M. A. Hardy (Ed.), *Studying aging and social change: Conceptual and methodological issues* (pp. 129–147). Thousand Oaks, CA: Sage.

Pavalko, E. K., & Elder, G. H. Jr. (1990). World War II and divorce: A life-course perspective. *American Journal of Sociology, 95*, 1213–1234.

Pittman, T. S., & Heller, J. F. (1987). Social motivation. *Annual Review of Psychology, 38*, 461–489.

Reiss, A. J. Jr., Duncan, O. D., Hatt, P. K., & North, C. C. (1961). Occupations and social status. New York: Arno.

Riesman, D., & Jencks, C. S. (1979). The visibility of the American college. In N. Sanford, & J. Axelrod (Eds.), *College and character* (pp. 40–62). Berkeley: Montaigne.

Riley, M. W., Johnson, M., & Foner, A. (1972). Aging and society: A sociology of age stratification. New York: Russell Sage Foundation.

Riley, M. W., & Riley, J. W. J. (1999). Sociological research on age: Legacy and challenge. *Ageing and Society, 19*, 123–132.

Rosenbaum, J. E. (1976). *Making inequality: The hidden curriculum of high school tracking*. New York: John Wiley and Sons.

Rosenberg, M. J., & Abelson, R. P. (1960). An analysis of cognitive balancing. In C. I. Hovland, & M. J. Rosenberg (Eds.), *Attitude organization and change: An analysis of consistency among attitude components*. New Haven, CT: Yale University Press.

Rosenberg, M. (1979). *Conceiving the self*. New York: Basic Books.

Rosenberg, M. (1985). Self-concept and psychological well-being in adolescence. In R. L. Leahy (Ed.), *Development of the self*. Orlando, FL: Academic Press.

Rosenberg, M. (1981). The self-concept: Social product and social force. In M. Rosenberg, & R. H. Turner (Eds.), *Social psychology: Sociological perspectives* (pp. 593–624). New York: Basic Books.

Rosenberg, M. (1965). *Society and the adolescent self-image*. Princeton, NJ: Princeton University Press.

Rosenberg, M., & Owens, T. J. (2001). Low self-esteem people: Collective portrait. In T. J. Owens, S. Stryker, & N. Goodman (Eds.), *Extending Self-Esteem Theory and Research: Sociological and Psychological Currents* (pp. 400–436). New York: Cambridge University Press.

Rosenberg, M., & Pearlin, L. I. (1978). Social class and self-esteem among children and adults. *American Journal of Sociology, 84*, 53–77.

Rosenberg, M., Schooler, C., & Schoenbach, C. (1989). Self-esteem and adolescent problems: Modeling reciprocal effects. *American Sociological Review, 54*, 1004–1018.

Rosenberg, M., Schooler, C., Schoenbach, C., & Rosenberg, F. (1995). Global self-esteem and specific self-esteem: Different concepts, different outcomes. *American Sociological Review, 60*, 141–156.

Rosenhan, D. L., Salovey, P., Karylowski, J., & Hargis, K. (1981). Emotion and altruism. In J. P. Rushton, & R. M. Sorrentino (Eds.), *Altruism and helping behavior* (pp. 21–38). Hillsdale, NJ: Erlbaum.

Rosow, I. (1974). Socialization to old age. Berkeley: University of California Press.

Rothbart, G., Sloan, L., & Joyce, K. (1981). Educational and work careers: Men in the Vietnam generation. U.S. Veterans Administration *Legacies of Vietnam: Comparative adjustment of veterans and their peers* Vol. Vol II. Washington D.C.: U.S. Government Printing Office.

Roy, A. D. (1983). Some thoughts on the distribution of earning. In G. S. Maddala (Ed.), *Limited-dependent and qualitative variables in econometrics* (pp. 135–146). Cambridge: Cambridge University Press.

Ryder, N. B. (1965). The cohort as a concept in the study of social change. *American Sociological Review, 30*, 843–861.

Sampson, R. J., & Laub, J. H. (1996). Socioeconomic achievement in the life course of disadvantaged men: Military service as a turning point, circa 1940–1965. *American Sociological Review, 61*, 347–367.

Santoli, A. (1981). *Everything we had: An oral history of the Vietnam War by thirty-three soldiers who fought it.* New York: Random House.

Scheff, T. J. (1987). Shame and conformity: The deference-emotion system. *American Sociological Review, 53,* 395–406.

Schwalbe, M. L., & Staples, C. L. (1991). Gender differences in sources of self-esteem. *Social Psychology Quarterly, 54,* 158–168.

Scott, J. F. (1971). *Internalization of norms.* Englewood Cliffs, NJ: Prentice Hall.

Sears, D. O. (1981). Life-stage effects on attitude change, especially among the elderly. In S. B. Kiesler, J. N. Morgan, & V. K. Oppenheimer (Eds.), *Aging: Social change* (pp. 183–204). New York: Academic Press.

Segal, D. R., & Wechsler, M. (1983). Change in military organization. *Annual Review of Sociology, 9,* 151–170.

Segal, D. R., & Wechsler, M. (1976). The impact of military service on trust in government, international attitudes, and social status. In N. L. Goldman, & D. R. Segal (Eds.), *The social psychology of military service* (pp. 201–211). Beverly Hills: Sage.

Settersten, R. A., & Owens, T. J. (eds). (2002). *New frontiers in socialization, vol. 7 Advances in Life Course Research.* Oxford: Elsevier Science.

Settersten, R. A. Jr. (2003). Age structuring and the rhythm of the life course. In J. T. Mortimer, & M. J. Shanahan (Eds.), *Handbook of the life course* (pp. 81–98). New York: Kluwer Academic/Plenum Publishers.

Settersten, R. A. Jr. (1999). *Lives in time and place: The problems and promise of developmental science.* Amityville, NY: Baywood.

Sewell, W. H., & Hauser, R. M. (1975). *Education, occupation, and earnings: Achievement in the early career.* New York: Academic Press.

Shapiro, D. (1983). Working youth. In M. E. Borus (Ed.), *Tomorrow's workers* (pp. 23–58). Lexington, MA: Lexington Books.

Sheppard, H. L. (1973). Youth discontent and the nature of work. In D. Gottlieb (Ed.), *Youth in contemporary society* (pp. 99–112). Beverly Hills: Sage.

Simmel, G. (1950). The sociology of Georg Simmel. K. H. Wolff (translated and edited by). New York: Free Press.

Simmons, R. G., & Blyth, D. A. (1987). *Moving into adolescence: The impact of pubertal change and school context.* New York: Aldine de Gruyter.

Spenner, K. I., & Featherman, D. L. (1978). Achievement ambitions. *Annual Reivew of Sociology, 4,* 373–420.

Spenner, K. I., & Featherman, D. L. (1978). Achievement ambitions. *Annual Review of Sociology, 4,* 373–420.

Staudinger, U. M., & Pasupathi, M. (2000). Life-span perspectives on self, personality, and social cognition. In F. I. M. Craik, & T. A. Salthouse (Eds.), *The handbook of aging and cognition* (2nd ed., pp. 633–688). Mahwah, NJ: Lawrence Erlbaum Associates.

Stouffer, S. A. et. al. (1949). *The American soldier: Adjustment during army life* Vol. 1 ed.. Princeton, NJ: University of Princeton Press.

Suchman, E. A., Williams, R. M., & Goldsen, R. K. (1953). Student reaction to impending military service. *American Sociological Review, 62,* 290–301.

Sudman, S. (1976). Applied sampling. New York: Academic Press.

Sullivan, H. S. (1953). Conceptions of modern psychiatry. With a foreword by the author and a critical appraisal of the theory by Patrick Mullahy. 2nd ed. New York: Norton.

Super, D. E. (1984). Career and life development. In D. Brown, & L. Brooks (Eds.), *Career choice and development: Applying contemporary theories to practice* (pp. 192–234). San Francisco: Jossey-Bass.

Super, D. E. (1970). Career development. In J. R. Davitz, & S. Ball (Eds.), *Psychology of the educational process* (pp. 428–475). New York: McGraw-Hill.

Super, D. E. (1957). *The psychology of careers: An introduction to vocational development*. New York: Harper Brothers.

Super, D. E., Starishevsky, R., Matlin, R., & Jordaan, J. P. (1963). *Career development: Self-concept theory*. New York: College Entrance Examination Board.

Swann, W. B. Jr., Stein-Seroussi, A., & Giesler, R. B. (1992). Why people self-verify. *Journal of Personality and Social Psychology, 62*, 392–401.

Taylor, S. E., & Brown, J. D. (1988). Illusion and well-being: A social psychological perspective on mental health. *Psychological Bulletin, 103*, 193–210.

Thornton, R., & Nardi, P. M. (1975). The dynamics of role acquisition. *American Journal of Sociology, 80*, 870–885.

U.S. Bureau of the Census. (1966). *Americans at Mid-Decade*. Washington, DC: Government Printing Office, P-23, No. 16.

U.S. Bureau of the Census. (1957). *Current Population Reports: Population Characteristics*. Washington, DC: Government Printing Office.

U.S. Bureau of the Census. (1956). *Marital and Family Status: March 1956*. Washington, DC: Government Printing Office.

U.S. Bureau of the Census. (1965). *Statistical Abstracts of the U.S.* Washington, DC: Government Printing Office.

U.S. Department of Commerce. (1967). *Special Report on Household Ownership and Purchases of Automobiles and Selected Household Durables, 1960–1967*. Washington, DC: Government Printing Office, P-65, No. 18.

U.S. Department of Labor. (2004). *The employment situation*. Web site: URL http://research.stlouisfed.org/fred2/data/UNRATE.txt

U.S. Veterans Administration. (1977). *Data on Vietnam Era Veterans*. Washington, D.C: Veterans Administration, Office of the Controller, Reports and Statistics Service.

Uhlenberg, P., & Mueller, M. (2003). Family context and individual well-being: Patterns and mechanisms in life course perspective. In J. T. Mortimer, & M. J. Shanahan (Eds.), *Handbook of the life course* (pp. 123–148). New York: Kluwer Academic/Plenum Publishers.

Van Maanen, J. (1976). Breaking in: Socialization to work. In R. Dubin (Ed.), *Handbook of work, organization, and society* (pp. 67–130). Chicago: Rand McNally College Publishing.

Van Maanen, J., & Schein, E. H. (1979). Toward a theory of organizational socialization. *Research in Organizational Behavior, 1*, 209–264.

Veblen, T. (1948). The theory of the leisure class. In M. Lerner (Ed.), *The portable Veblen*. New York: The Viking Press.

von Eye, A., Kreppner, K., Spiel, C., & We-Sels, H. (1995). Life events' spacing and order in individual development. In T. A. Kindermann, & J. Valsiner (Eds.), *Development of person-context relations* (pp. 147–164). Hillsdale, N.J.: L. Erlbaum Associates.

Wallace, W. L. (1988). Toward a disciplinary matrix in sociology. In N. J. Smelser (Ed.), *Handbook of sociology* (pp. 23–76). Newbury Park, CA: Sage.

Weber, M. (1978). *Economy and Society: Vol 1 & 2*. Berkeley: University of California.

Weisberg, S. (1985). *Applied linear regression* 2nd ed.. New York: John Wiley and Sons.

Wells, L. E., & Marwell, G. (1976). *Self-esteem: Its conceptualization and measurement*. Beverly Hills: Sage.

Wethington, E. (2002). The Relationship of Turning Points at Work to Perceptions of Psychological Growth and Change. In R. A. Settersten, & T. J. Owens (Eds.), *New Frontiers in socialization* (Vol. 7 pp. 111–131). Oxford, Eng: Elsevier Science.

Wheaton, B. (1987). Assessment of fit in overidentified models with latent variables. *Sociological Methods and Research, 16*, 118–154.

Wheaton, B., Muthen, B., Alwin, D. F., & Summers, G. F. (1977). Assessing reliability and validity in panel models. In D. Heise (Ed.), *Sociological Methodology 1977* (pp. 84–136). San Francisco, CA: Jossey-Bass.

Wheeler, S. (1966). The structure of formally organized socialization settings. In O. G. Jr. Brim, & S. Wheeler (Eds.), *Socialization after Childhood: Two essays*. New York: John Wiley and Sons.

Whyte, W. F. (1981). *Street corner society: The social structure of an Italian slum* 3rd ed.. Chicago: University of Chicago.

Williamson, O. E. (1975). *Markets and hierarchies, analysis and antitrust implications: A study of the economics of internal organization*. New York: Free Press.

Winship, C., & Radbill, L. (1994). Sampling weights and regression analysis. *Sociological Methods & Research, 22*, 230–57.

Wrong, D. (1961). The oversocialized conception of man in modern sociology. *American Sociological Review, 26*, 184–193.

Wylie, R. C. (1974). *The self-concept: A review of methodological considerations and measurement instruments* rev ed., Vol. 1. Lincoln, NE: University of Nebraska.

Wylie, R. C. (1979). *The self-concept: Theory and research on selected topics* (rev ed.). Lincoln, NE: University of Nebraska Press.

# Index